MOSSES
LIVERWORTS
& HORNWORTS
OF THE WORLD

MOSSES
LIVERWORTS
& HORNWORTS
OF THE WORLD

A GUIDE TO
EVERY ORDER

Joanna Wilbraham

PRINCETON UNIVERSITY PRESS
PRINCETON AND OXFORD

Published in 2025 by Princeton University Press
41 William Street, Princeton, New Jersey 08540
99 Banbury Road, Oxford OX2 6JX
press.princeton.edu

Conceived, designed, and produced by
The Bright Press
an imprint of The Quarto Group
1 Triptych Place, London, SE1 9SH, United Kingdom
T (0) 20 7700 6700
www.quarto.com

GPSR Authorized Representative: Easy Access
System Europe - Mustamäe tee 50, 10621
Tallinn, Estonia, gpsr.requests@easproject.com

Library of Congress Control Number: 2025931444
ISBN: 978-0-691-26519-3
Ebook ISBN: 978-0-691-26877-4
British Library Cataloging-in-Publication Data is available

Publisher **James Evans**
Editorial Director **Isheeta Mustafi**
Art Director **James Lawrence**
Editor **Nick Pierce**
Project Manager **Ruth Patrick**
Design **Kevin Knight**
Picture Research **Katie Greenwood**
Illustrations **John Woodcock**

Page 2 photo: Hummocks of *Sphagnum* mosses thrive in the
extensive bogs of Rannoch Moor, Scotland, UK.

Cover photos: Front cover (clockwise from top left): *Plagiochila
appalachiana, Anthoceros fusiformis, Blasia pusilla, Myurium
hochstetteri, Oedipodium griffithianum, Discelium nudum,
Bryum argenteum, Conocephalum conicum, Encalypta ciliata,
Pleurozia gigantea.* Back cover: *Bryum.* Spine: *Rhodobryum
umbraculum.* All cover images by Des Callaghan, apart from:
Štepán Koval: front cover (bottom, centre) Shutterstock/Mykola
Borduzhak: back cover

Printed in Malaysia

10 9 8 7 6 5 4 3 2 1

MIX
Paper | Supporting
responsible forestry
FSC® C007207
FSC
www.fsc.org

CONTENTS

INTRODUCTION

ABOVE | From a view down at moss level, the capsules of these mosses, *Meesia hexasticha* and *M. uliginosa*, tower skyward on vibrant red setae.

OPPOSITE | Small green shoots of the tropical moss *Schlotheimia ferruginea* with abundant, striking large capsules.

Mosses, and their close relatives the liverworts and hornworts, are a highly diverse and successful group of plants that have adapted to thrive in nearly all parts of the planet. They can dominate the landscape entirely across Arctic tundras and they can survive in harsh dry deserts under conditions few other plants can tolerate. You can even find them a few steps from your home, growing happily in sidewalk cracks and forming soft green cushions along the top of walls.

They are plants with ancient origins. Their shared ancestry stretches back over 400 million years, when tectonic plates carried them across the earth's surface and mountains rose and fell. During this time, these small, green, and seemingly insignificant plants have become an integral part of many ecosystems. Mosses and liverworts

play an important role in water cycling. Carpeting tree branches and forest floors, they act as natural sponges to absorb rainwater and then release it at a more manageable rate into the surrounding habitat. *Sphagnum* mosses, and the peatlands that they create, lock carbon underground and play a crucial part in efforts to turn the tide on global warming.

In this rapidly changing world, bryophytes are under threat. They are particularly at risk from the rapid warming of the planet, as many species require cool, humid conditions. Understanding and appreciating their diversity and fascinating evolutionary history is an essential step toward ensuring this extraordinary group of plants is protected into the future.

WHAT ARE BRYOPHYTES?

The term "bryophytes" describes plants of three closely related evolutionary lineages; the Bryophyta (mosses), Marchantiophyta (liverworts), and Anthocerotophyta (hornworts). These three groups share several important characteristics in their evolution, life cycle, and anatomy that unite them. There are an estimated 20,000 species of bryophytes. The most species-diverse group is the mosses with *c.*12,500 species, followed by the liverworts with *c.*7,000. In comparison, the hornworts are by far the smallest division with just *c.*200 species.

Bryophytes are an ancient group of land plants that disperse by spores rather than seeds. They are small, easily overlooked plants. The inexperienced observer may view them as miniature, under-developed versions of the more familiar larger plants such as ferns, herbs, and trees; primitive precursors that have failed to pass the evolutionary finish line. However, this outlook would be missing a great deal. To understand bryophytes, we need to view the world at their scale. Bryophytes may be small in stature, but they are successful plants that are found on every continent and occupy a diverse array of habitats.

EVOLUTIONARY ORIGINS

The evolution of early land plants from their aquatic origins to life on land required some major innovations, one of which being the need to keep cells hydrated so they can remain metabolically active. Plants such as trees have solved this problem by evolving vascular tissue (the thick-walled tubular cells of xylem and phloem) to transport water and nutrients from the soil surrounding their roots up to the tip of the plant shoots. Rather than develop complex plumbing systems, bryophytes have followed an alternative strategy. They are "poikilohydric"—the water content inside the plant is directly affected by the water content on the outside.

Most bryophytes grow no more than a few centimeters high, carpeting forest floors or clinging to rocks and tree trunks. They live in the interface between the earth and the atmosphere, a boundary layer that is protected from winds and traps humidity. They have evolved desiccation tolerance, enabling them to suspend metabolic activity in periods of drought, to photosynthesize and grow again when water becomes available.

Instead of conducting fluids internally, bryophytes absorb water and nutrients externally across the whole body of the plant. Their body

plan is therefore restricted in size, since smaller shapes have a higher surface area relative to their internal spaces, and all living cells within the plant require direct access to the liquids and gases absorbed at the plant's surface. These strategies have not only enabled bryophytes to survive, but to thrive, perfectly adapted to a life in miniature.

ABOVE | The Laurisilva of Madeira is a rare temperate rainforest where high humidity enables bryophytes to flourish, forming thick carpets over the surface of the trees.

RIGHT | To the left, *Hedwigia stellata* is shown hydrated with spreading leaves. To the right, the dry shoots are pressed close to the stem, emphasizing the white-tipped leaves.

EARLY EVOLUTION

The mosses, liverworts, and hornworts are an ancient group of plants. They have their origins over 473 MYA in the Late Cambrian to Early Ordovician, when extensive, warm, shallow seas were bursting with a diversity of marine invertebrate life but the barren earth was only in the very early stages of colonization. Bryophytes are a monophyletic group, which means they share a common ancestor in their evolutionary history. They are the closest living relatives to the now extinct green algal lineage that first made the transition from aquatic to terrestrial life. This means that studying their biology provides fascinating clues to the evolution of early land plants.

The fossil record for bryophytes is poor compared to vascular plants such as ferns and seed plants, since they lack structures that preserve well such as lignified woody tissues. However, recent advances in palaeobotanical research suggest that to some extent bryophyte fossil evidence may also have been overlooked in the past and that the fossil record is in fact more informative than previously realized. We are gaining increasing clarity on the origins and relationships of these early land plants thanks to new fossil finds and advanced analytical techniques applied to modern genetic data, allowing us to estimate a timeline for when different lineages diverged.

The latest research suggests that the hornworts diverged from the rest of the ancestral bryophyte lineage 479–450 MYA in the Ordovician, the liverworts 440–412 MYA, and the mosses 420–364 MYA. The first evidence of bryophyte-like ancestors in the fossil record dates from the Ordovician period some 450 MYA and is based on distinctive fossil spores that resemble those of modern-day liverworts. Liverworts are more confidently present in the fossil record by the Devonian, where more complete fossils of recognizable plants have been found, although these finds have been very rare. In the first two decades of the twenty-first century, however, new and exciting fossil discoveries have been reported, including the earliest known bryophyte macro-fossil—a simple thalloid liverwort, *Riccardiothallus devonicus*, dated from the Lower Devonian (407–411 MYA). Each new fossil discovery can provide more clues about the evolution and diversification of these early plants, and these recent finds suggest there is still much more to discover.

By the end of the Cretaceous around 66 MYA, most fossils of mosses and liverworts can be identified as modern genera, and the bryophytes trampled by a marauding *Tyrannosaurus rex* would have a strong similarity to modern floras. The Late Cretaceous saw an increase in the angiosperms with the rise of the trees, creating a new habitat potential for bryophytes. A transition among some groups to a new lifestyle living on the trees for support and exploiting these new living structures was likely a major driver for innovation and speciation, with the creeping pleurocarpous mosses diversifying at around the same time period as trees became more commonplace. The trees also provided a new source of fossil evidence, since their sticky resin could trap fragments of plants and insects before hardening and ultimately becoming amber deposits containing some of the most beautiful and well-preserved bryophyte fossils.

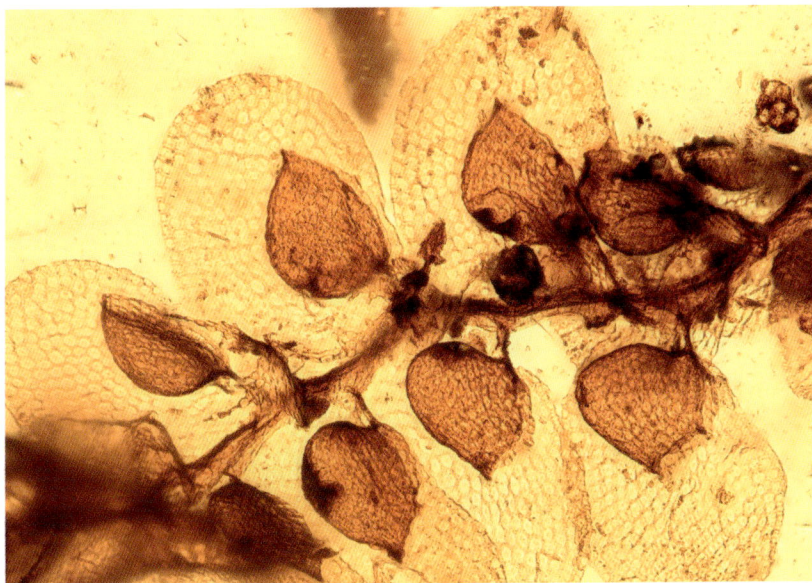

LEFT | This beautifully preserved shoot of *Frullania kachinensis* has been fossilized in amber, with even the cell structure of the leaves clearly visible.

BELOW LEFT | This fossil, named *Calymperites burmensis*, was discovered preserved in amber from the Cretaceous period and is thought to be a relative of the modern moss family Calymperaceae.

BELOW RIGHT | This liverwort *Protofrullania cornigera* is preserved in amber, allowing detailed examination of this ancient flora.

BRYOPHYTE LIFE CYCLE

Mosses, liverworts, and hornworts all share a similar life cycle, which is an important feature uniting these three groups as bryophytes.

The bryophyte life cycle is based on the alternation of two multicellular generations where a green, free-living, and often leafy phase known as the gametophyte alternates with a spore-producing phase, the sporophyte, which remains attached to the gametophyte, on which it is partially to entirely dependent.

Spores are produced through the cell division process of meiosis, meaning the spore cells contain only a single set of chromosomes (haploid). When the spores germinate, a juvenile stage of algae-like filaments known as the protonema grows. This fuzzy green mat extends over the growing surface and can spread over several centimeters. This is usually a very short-lived stage in the plant's life cycle, and from this protonema one or more leafy plants usually arise. There are exceptions, though, for example in the Southeast Asian moss *Ephemeropsis tjibodensis*, where the reproductive structures and sporophytes are produced directly on a persistent protonema with no leafy shoots produced at all.

This dominant leafy (or thallose) stage in the life cycle, the gametophyte, is also haploid. This is the phase of the plant that produces the gametes or sex organs: archegonia (female) and antheridia (male). These may be produced on the same plant (monoicous) or separate plants (dioicous). The archegonia are flask-shaped and consist of an enlarged basal portion where the female egg cell (or gamete) is located and an elongated upper portion through which the male gametes (or sperm cells) must swim to reach the egg cell. To help matters along, the archegonium releases a chemical attractant so the sperm cells set off in the right direction. At fertilization, the egg and sperm cells unite and form a cell known as a zygote containing paired chromosomes. This is the beginning of the diploid generation that grows into the sporophyte. The developing sporophyte is anchored to the gametophyte tissue, absorbing its nutrients. In most bryophytes the sporophyte develops a stalk-like seta with a capsule at the top from which the spores are produced and subsequently dispersed, ready to begin the cycle again.

VEGETATIVE REPRODUCTION

Sexual reproduction is a costly endeavor requiring a huge reserve of energy and resources. For some bryophytes sex is not even an option. A study of British moss species found that around 14 percent never undergo sexual reproduction at all. Dioicous species in particular can struggle to reproduce sexually as both male and female plants must be in close proximity. For the liverwort *Plagiochila exigua*, egg and sperm cells are separated by the Atlantic Ocean, since only female plants appear to occur in North America while only males are found in Europe. Fortunately, bryophytes have developed a variety of mechanisms with which they can reproduce by cloning small propagules of themselves. Some of these vegetative or asexual reproduction strategies are as simple as having the ability to regrow from discarded plant fragments. There are also special structures such as gemmae or tubers that act as specialized vegetative propagules.

THE MOSS LIFE CYCLE

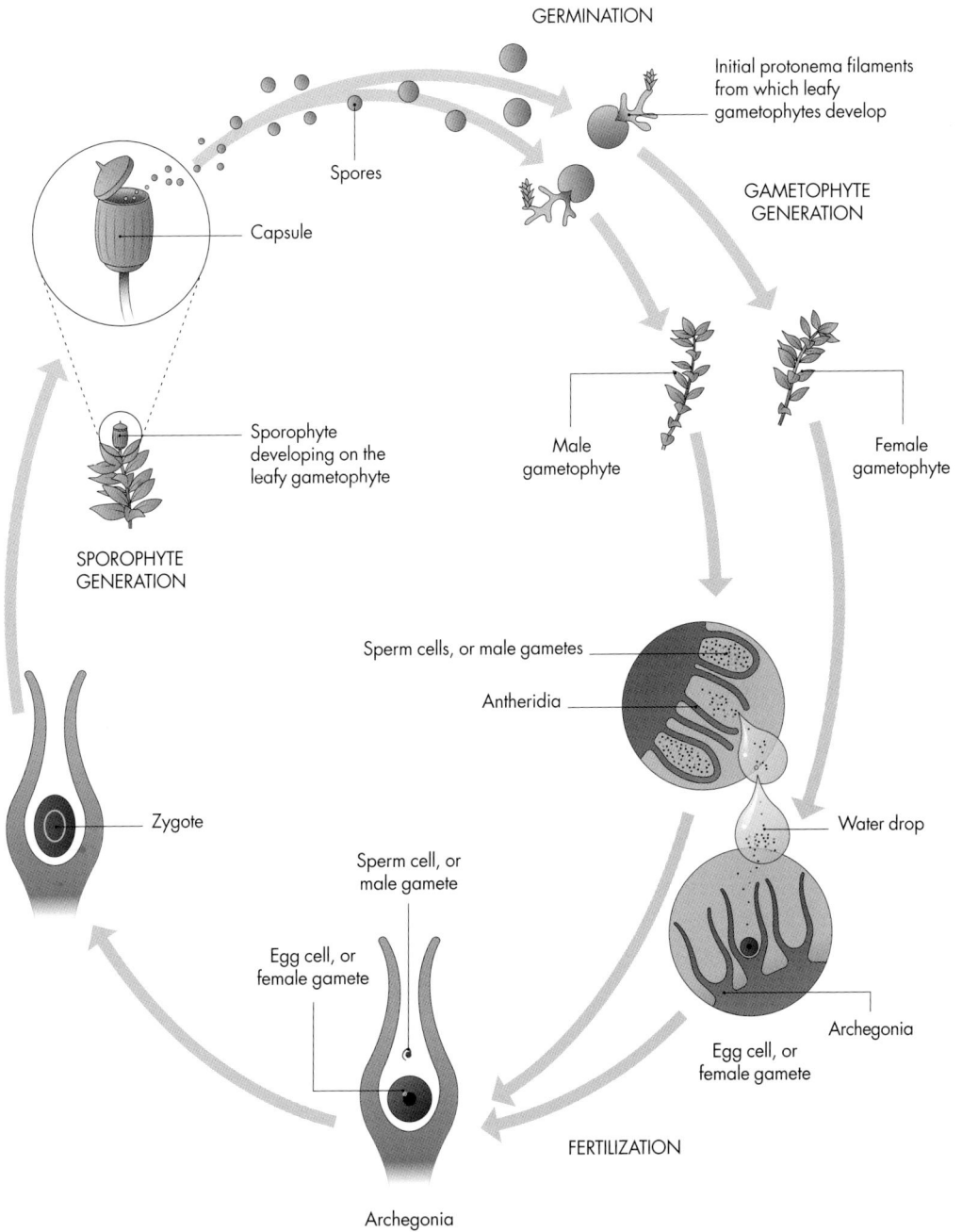

GERMINATION

Spores

Capsule

Initial protonema filaments
from which leafy
gametophytes develop

GAMETOPHYTE
GENERATION

Sporophyte
developing on the
leafy gametophyte

Male
gametophyte

Female
gametophyte

SPOROPHYTE
GENERATION

Sperm cells, or male gametes

Antheridia

Water drop

Zygote

Sperm cell, or
male gamete

Egg cell, or
female gamete

Archegonia

Egg cell, or
female gamete

FERTILIZATION

Archegonia

BRYOPHYTE ANATOMY

For seemingly simple plants the bryophytes have a surprisingly diverse array of anatomical features. Because bryophytes lack vascular tissue, they do not have true leaves and stems as in flowering plants but for convenience the leaf and stem-like structures are referred to by these terms. This section is not intended to be an exhaustive treatment of all the fascinating diversity of bryophyte anatomy but should provide the background needed to interpret the taxonomic profiles.

IS IT A MOSS, A LIVERWORT, OR A HORNWORT?

Upon close inspection the three groups of bryophytes are easily distinguished using a combination of features. The difficulty comes in ascribing an absolute morphological distinction between them as many features overlap. The mosses almost always have their leaves arranged spirally around the stem, the hornworts and some liverworts form thalloid green plates, while the leafy liverworts have their leaves in two or three rows. For those plants that produce a leaf-like structure, moss leaves are formed undivided and may have a line of thickened tissue called the costa running up the center of the leaf, while the leafy liverworts never have a costa but can form leaves that are divided or lobed.

The most obvious distinctions can be seen when the bryophytes are producing their sporophytes. The moss sporophyte is usually a robust structure; the seta is rigid and elongates before the capsule spores are mature, and can persist for some time after the spores are released. The liverwort seta, in contrast, is a more ephemeral affair, which elongates quickly

after the spores mature and then decays soon after the spores are released. Hornworts are easily recognized when their distinctive capsules are present, though sterile specimens could be confused with thalloid liverworts.

The mechanism of spore dispersal is associated with different specialist structures across the different groups of bryophytes. All three groups make use of hygroscopic movement to disperse spores. Liverworts and hornworts have small spring-like structures in the capsule that help break up the spore mass. Mosses often have a sophisticated structure called the peristome surrounding the capsule opening, which can open and close depending on humidity.

HORNWORTS

Hornworts have the most uniform structures of the three bryophyte groups, though their outward simplicity disguises some interesting complexities at microscopic level.

The vegetative gametophyte

The vegetative gametophyte is a flattened green thallus that typically forms rosettes or ribbons. The thallus lacks internal differentiation, though most genera have intercellular spaces that exude mucilage. The thallus surface may also have mucilage clefts that consists of two kidney-shaped cells surrounding an open pore, and these function as a site of entry for the *Nostoc* cyanobacteria that is found as an endosymbiont in all hornworts. The underside of the thallus produces mostly unbranched rhizoids, which anchor the plant to the substrate.

Spores

The sporophyte capsule

Involucre

Flattened thallus
of the gametophyte

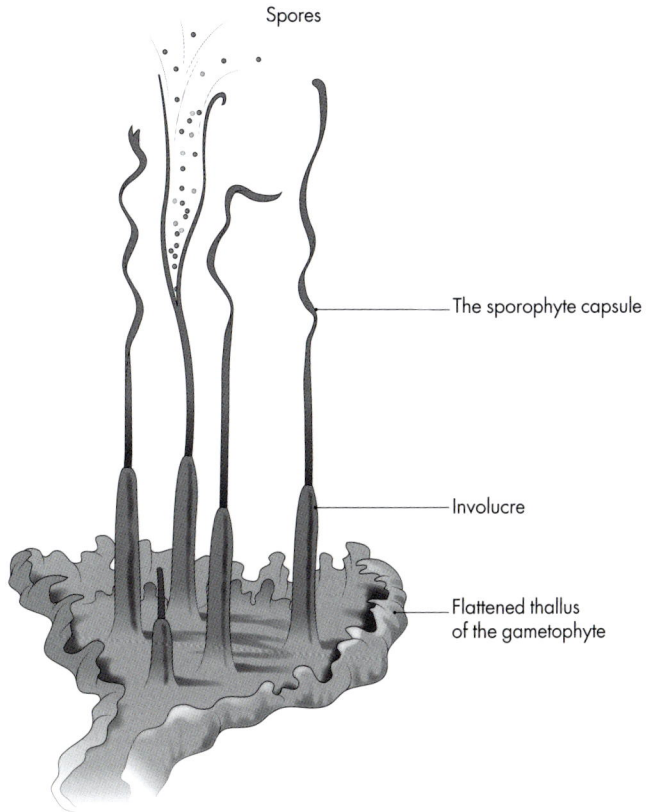

At the cellular level hornworts share some commonality with the algae. Their cells usually have just one solitary chloroplast, and this often contains a structure called a pyrenoid, which has a role in concentrating carbon dioxide as part of their photosynthetic processes.

Reproductive structures

The female sex organs (the archegonia) develop from cells on the thallus surface and sink into the thallus as they develop. The archegonia are shaped

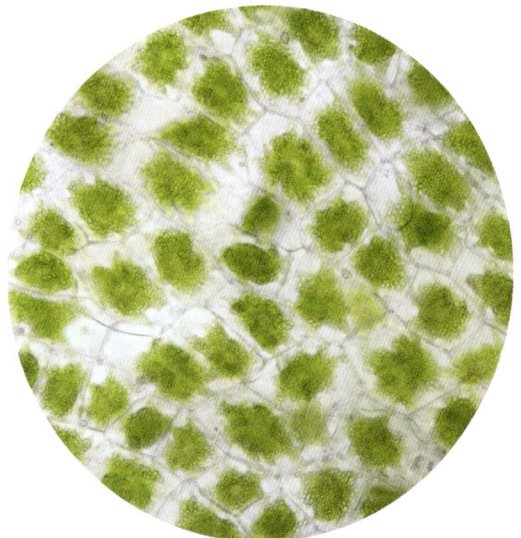

RIGHT | Hornwort cells can be recognized under the microscope when there is just one green chloroplast in each cell, as in this *Anthoceros agrestis*.

like a vase with a short neck and hold a single egg. They protrude slightly above the surface as they mature, where they are protected by mucilage and a few cells. The male sex organs, the antheridia, develop beneath the thallus surface in an internal chamber where, depending on the genus, 1–80 antheridia may develop. When the antheridia are mature, the roof of this chamber ruptures and the sperm cells can swim out. Asexual reproduction is also widely employed by hornworts. They can propagate from detached fragments of thallus, and some genera produce tubers or the vegetative buds known as gemmae.

Spore-producing structures

Following fertilization the newly formed zygote is divided into a basal region that evolves into a foot and an upper region that develops into the capsule. As the young sporophyte is developing, it is enclosed by protective gametophyte tissue called the involucre, and this persists when the capsule is mature as a cylinder at its base. The foot embeds into the gametophyte tissue and forms an interface between the two generations by which the sporophyte can obtain nutrients. The capsule grows into an elongated cylindrical structure in which spores are produced continually through the growing season. The capsule also contains structures called pseudoelaters, which break up the spore mass. The spores are released by two slits that open up in longitudinal lines down the capsule. Hornworts show a wide diversity of spore ornamentation and color, and these details are often needed to identify species.

LIVERWORTS

Liverworts have an extreme range of morphologies that can be broadly grouped into three main categories: the simple and complex thalloid liverworts and the leafy liverworts.

The vegetative gametophyte

The simple thalloid liverworts have an undifferentiated thallus, which may only be one cell thick in some genera, for example the Metzgeriales. In the complex thalloid liverworts the thallus is several cell layers thick and is internally differentiated with air chambers and layers of green photosynthetic tissue and storage cells. Rows of scales may also form underneath the thallus. Some complex thalloids produce two kinds of rhizoids: so-called "pegged" rhizoids with cell wall projections into the interior of the cell that are involved in water transport, and smooth rhizoids that anchor the plant and also facilitate access to the plant by endophytic fungi.

The leafy liverworts have stems and leaves, usually with two lateral rows of leaves and often a third row of smaller leaves on the underside of the stem. The lateral leaves of some species are divided into two lobes, the smaller of which can be modified into a water sac. The precise orientation by which the leaves are positioned on the stem is an important feature in leafy liverwort taxonomy. While the underleaves are positioned horizontally on the stem, the lateral leaves have an angled attachment which has some significant variations. If the upper edge of each leaf overlaps the lower margin of the leaf above it on the stem, then this is known as an incubous leaf arrangement. A succubous arrangement would be when the upper margin of the leaf is hidden under the lower margin of the leaf above. Liverwort leaves develop from up to three dividing growth points, which means the leaves can be very elaborate—lobed or finely segmented.

Within the leaf cells, liverworts can produce unique structures called oil bodies, which are absent in both mosses and hornworts. Their

The undifferentiated thallus of a typical simple thalloid

SIMPLE THALLOID LIVERWORT

Gemmae cup

Female gametophyte with archegoniophores

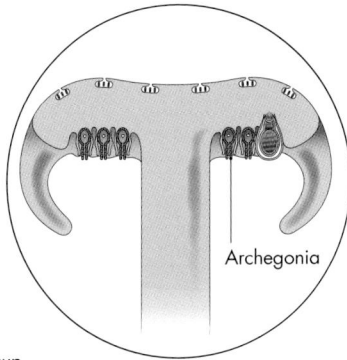

Archegonia

Female receptacle

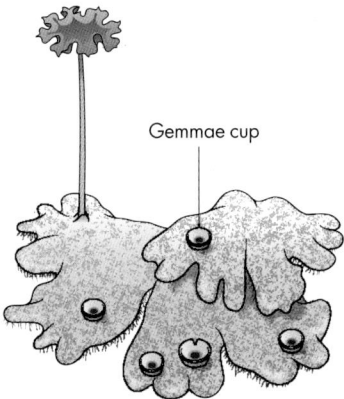

Gemmae cup

Male gametophyte with antheridiophore

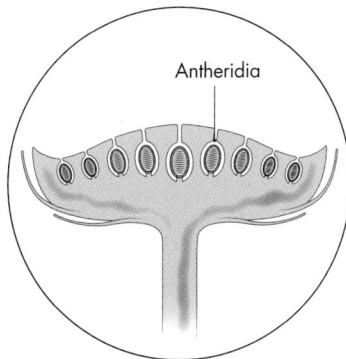

Antheridia

Male receptacle

COMPLEX THALLOID LIVERWORT

LEAFY LIVERWORTS

The upperside of
a shoot showing
an incubous leaf
arrangement

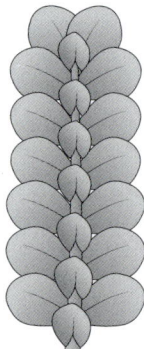

The underside of
a shoot with two rows
of leaves and a row
of underleaves

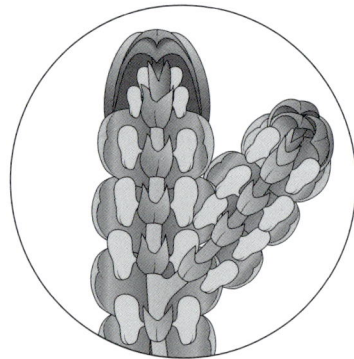

Leafy liverwort leaves
can form complex lobes
with some functioning
as water sacs

function is not certain but may be involved in deterring herbivores or possibly in protection from UV radiation.

Reproductive structures

The simple thalloid liverworts have archegonia and antheridia that develop on the thallus surface, clustered in sunken chambers or protected by scales of tissue. In some complex thalloids the antheridia and archegonia develop on the surface of receptacles that are raised up on specialist structures known as gametophores, where the sporophytes develop on the umbrella-like structures.

In leafy liverworts the antheridia develop in the axil of specialized leaves either on the main shoot or specialized male branches. The archegonia develop in clusters at the apex of a branch and are protected by leaves that fuse together to form a perianth. In some thalloid liverworts the sporophytes are surrounded by a tubular involucre, similar to that in hornworts. Some species develop two involucres, in which case the innermost one is called the pseudo-perianth, and this resembles the perianth of leafy liverworts. Vegetative reproduction is common in liverworts, with propagation from gemmae, tubers, and plant fragments widespread across the group.

Spore-producing structures

The liverwort sporophyte is organized into a foot, a seta, and a capsule. The translucent seta elevates the capsule once the spores have matured. The spore-producing tissue inside the capsule develops both spores and spiraled elaters to disperse the spore sac. Most liverwort capsules divide into four valves that reflex backward on drying and allow the spores to escape.

OPPOSITE TOP | The complex structures of this *Frullania tamarisci* plant can be clearly seen in this scanning electron microscope image.

MOSSES

The majority of the mosses can be divided into two groups: the acrocarpous and the pleurocarpous mosses, which are defined by the position of the sporophytes on the plant. Acrocarpous mosses have simple upright shoots that are sparsely branched, with the female reproductive organs produced at the tips of the main shoots, terminating further growth. For example, the Pottiales are acrocarpous mosses that typically form small, neat cushions. Pleurocarpous mosses have a creeping shoot system, and sporophytes are produced at the tips of small side branches coming off the main stem, such as in the Hypnales order. There are some genera that have different architectural branching patterns and do not quite fit either of these concepts, but the categorization of acrocarpous and pleurocarpous for the majority of mosses is a useful distinction.

The vegetative gametophyte

Leaves in mosses mostly have a spiral arrangement on the stem. Moss leaves do not vary a great deal in shape compared to the liverworts, since they develop from only one growing point and are typically egg-shaped to narrowly tapering. The leaf lamina is usually one cell thick, though in some species has multiple layers, often with a thicker costa region running up the middle. The function of the costa is mainly to provide physical support to the leaf, though it also provides some limited transport of water and nutrients.

The pattern of cells in the moss leaf provides a lot of useful taxonomic information. The cells can be elongate to rounded; the leaf edge may be differentiated, forming a border; and the cells may sometimes protrude to create a toothed margin. Some mosses have specialized, thick-walled, and sometimes inflated alar cells in the basal corners of the leaf, which have a role in optimally

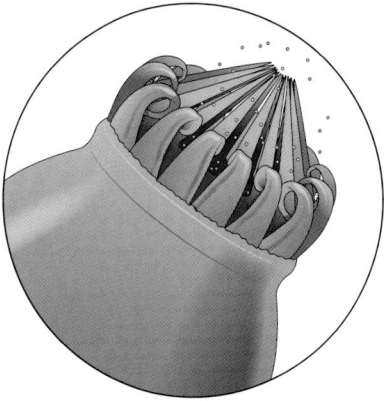

A MOSS CAPSULE WITH
A DOUBLE PERISTOME

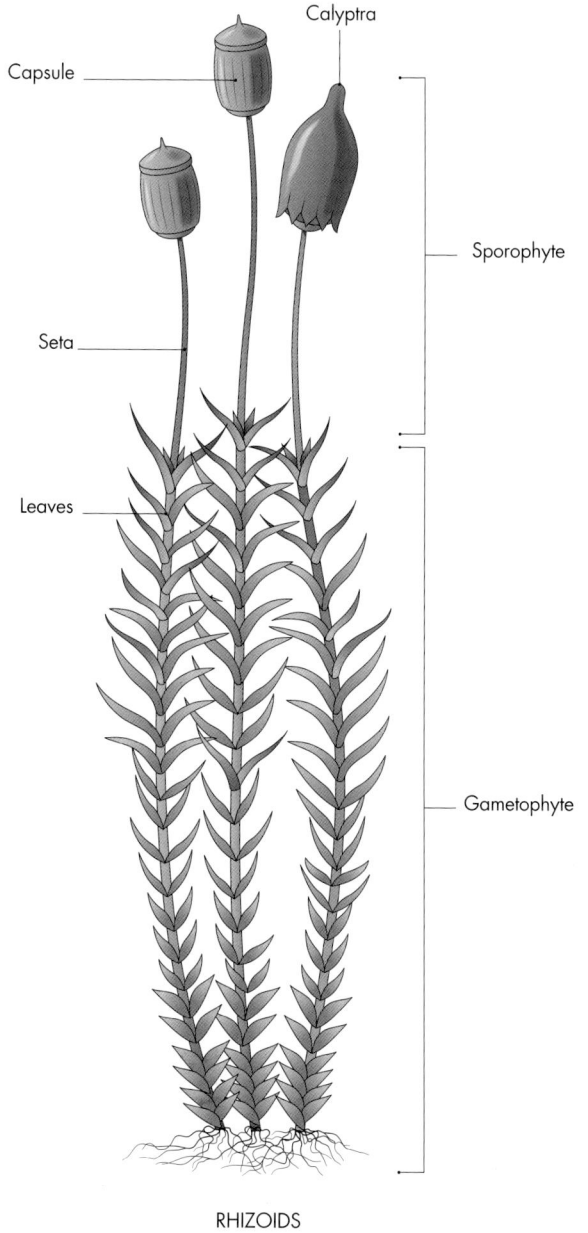

Calyptra

Capsule

Sporophyte

Seta

Leaves

Gametophyte

ACROCARPOUS MOSS

RHIZOIDS

PLEUROCARPOUS MOSS

positioning the leaf orientation. The cell walls of the leaf can be smooth or variously ornamented with papillae. Multicellular rhizoids serve to anchor the plants to the substrate. In some mosses, densely packed rhizoids can form a felt-like mat on the stem, and in these cases they likely contribute to the movement of water by capillary action along the outer surface of the stem.

Reproductive structures

Antheridia and archegonia appear at the apex of branches, forming terminal clusters that are protected by specialist leaves. The clusters of antheridia and their leaves are called perigonia, and the female equivalent are the perichaetia. In some mosses the leaves around the antheridia form a splash cup to increase the dispersal distance when raindrops hit and splash out the sperm cells. These perigonial and perichaetial leaves are usually different shapes and sizes to the vegetative leaves and provide useful taxonomic characters.

Like the rest of the bryophytes, mosses make use of a variety of mechanisms to clonally propagate themselves. As well as producing gemmae and tubers, some mosses have specific small branches and leaf bundles that become easily detached and act as propagules.

Spore-producing structures

The moss sporophyte comprises the foot, which anchors the seta into the gametophyte tissue, and a capsule, which is usually a lot more complex than in liverworts. Most mosses disperse their spores via a capsule lid called an operculum, which falls away when the spores are mature. Many mosses also have a peristome, which is a sophisticated structure to facilitate spore dispersal. It is formed of a circular system of teeth inserted into the opening of the capsule, and there are significant variations in how these teeth are formed and how they are arranged. The inner and outer cell walls of the peristome absorb water to different extents, causing the teeth to bend over when wet, opening up the capsule mouth for spore release.

Unique to mosses is the little hat called the calyptra that covers the capsule. This is actually composed of the gametophyte tissue and forms as the sporophyte is developing. There are generally two main forms of calyptra: hood-shaped, where the "hat" is split up one side, and cone-shaped, which may also be lobed at the base. The calyptra has a role in the development of the capsules, and if it is experimentally removed at an early stage, the capsule will become misshapen. It also has a role in preventing the young sporophyte from drying out.

RIGHT | The sporophyte of the leafy liverwort *Radula complanata* develops enclosed within the perianth tube. The mature capsule is elevated on a seta and then releases its spores by splitting into four valves.

GEOGRAPHICAL DISTRIBUTIONS AND HABITATS

Many bryophytes have wide-ranging distributions, often spanning more than one continent. They have a great capacity for dispersal, with their windborne spores having the potential to travel great distances. However, within these broad ranges, species often have very particular environmental requirements and are found only in certain kinds of habitats or on a particular kind of surface.

BELOW | The thalloid liverwort *Conocephalum conicum* has spore-bearing structures elevated up high on exceedingly tall setae.

OPPOSITE | A cloud of spores can be seen escaping from the capsule of a *Polytrichum* moss.

BRYOPHYTE DISTRIBUTIONS

The ancient origins of bryophytes mean these plants have been around for a very long time, and so we also see patterns in their distribution that reflect geological history. Some bryophyte species are found in different regions of the world separated by considerable distances, and these are known as disjunct distributions. It can be challenging to interpret whether these patterns reflect long-distance dispersal events or whether they originate from continental drift processes responsible for separating land masses and thrusting up mountain ranges. There are likely

to have been a whole spectrum of past events resulting in the picture of bryophyte distribution we have currently. Changes in past climate or habitat conditions can also dramatically influence distributions, and we see examples today of bryophytes that have managed to persist as relict populations within a larger landscape that in the past became largely uninhabitable—for example in the extensive glaciation of the northern hemisphere where small, sheltered pockets of habitat escaped the ice and allowed species to persist.

A better understanding of species diversity based on genetic analysis has also helped identify more so-called "cryptic" species. This is where a species that could be quite widespread is found to actually consist of more than one genetically distinct entity. Often, small yet consistent features in the plant's appearance that were previously considered insignificant can now be used to distinguish these species once the genetic groupings have been identified. The large thalloid liverwort *Conocephalum conicum* was thought to range across North America, Europe, and East Asia. Recent studies, however, have found that several distinct species were actually included under this concept of *C. conicum*. Most of these newly described cryptic species have a more restricted distribution range within the footprint of what had previously been considered one broadly defined species.

In more recent times, human activities have unwittingly introduced certain bryophyte species into different parts of the world, where they have become established. The moss *Campylopus*

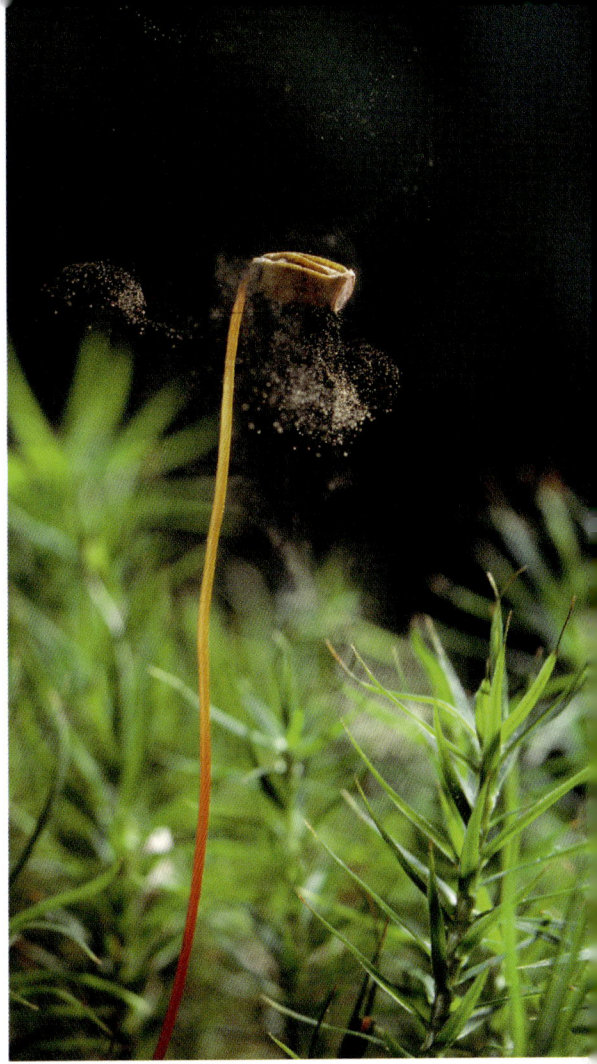

introflexus originates from the southern hemisphere and was first recorded as an introduction to Britain in 1941. By 1990 it was widespread and established across the country. Assessing the impact of introduced species on the native flora is challenging in the face of accelerating changes in the environment. What is clear is that bryophyte distributions are dynamic, with ranges extending but unfortunately more commonly contracting. There has never been a more urgent time to document species diversity and monitor their distributions to provide data for conservation management strategies in a changing world.

BRYOPHYTE HABITATS

Bryophytes are found in a surprisingly diverse range of habitats. It is important to remember the scale at which these small plants interact with their environment. Even a seemingly inhospitable landscape can often be found to harbor some moss or liverwort diversity where small pockets of more suitable habitat are found. Some species can be restricted to very particular habitats, be it alpine snowbeds, chalk grassland, or rotten logs, whereas others are tolerant enough to be found on a range of substrates. Sometimes unexpected anomalies occur—for example, the magnificently handsome moss *Myurium hochstetteri* is restricted to only two regions of the world. In Britain it grows exclusively on rocks or slopes near the sea in the far reaches of northwest Scotland. It is otherwise found only in the Macaronesian islands of Azores and Madeira, where it is, in contrast to its Scottish habitats, a common forest epiphyte.

HUMAN-MADE ENVIRONMENTS

You do not need travel to exotic destinations to find bryophytes. There are many pioneer species adapted to harsh environments, and urban specialists include a suite of stress-tolerant and adaptable species. Cities are an interesting place to study changes in the local environment, and bryophytes can be useful indicators of pollution. Where clean-air policies have been implemented in recent decades, a reduction in sulfur dioxide emissions has resulted in a return of mosses after long periods when city air may have been too polluted. In London this is particularly noticeable with the recent increase of epiphytic moss species such as *Orthotrichum*, which form small cushions on

trees. These examples go to show that there can be space for bryophytes in cities and that what is good for bryophytes is good for people too—both share a need for green spaces and clean air.

People have been modifying the natural landscape for thousands of years, and while these impacts have been largely detrimental to biodiversity in terms of habitat loss, in some instances such activities have created opportunities for bryophytes. Arable fields in Europe have been traditionally managed by annual plowing with the stubble fields left uncultivated over winter following the harvest. There are several rare hornworts and liverworts that can be found almost exclusively in arable fields since they are adapted to colonizing bare soil and cannot cope with competition for space. Unfortunately, these habitats are under threat from more intensive agricultural methods.

A surprising habitat that supports some exceedingly rare species is old abandoned mining sites. There is a small group of bryophytes known as metallophytes that have a high tolerance for heavy metals toxic to other species. The so-called "copper mosses," *Mielichhoferia mielichhoferiana* and *Scopelophila cataractae*, for example, occur only on rocks rich in copper sulfide or soils containing this heavy metal. Mining activities have brought an increased surface area of affected soils to the surface, available for colonization. It is likely that these mosses gain some advantage in being able to grow in these seemingly toxic conditions in which competitors would not be able to thrive.

OPPOSITE | Even on this busy London street, a diversity of bryophytes can be found on a wall top.

BELOW LEFT | A close-up of the *Mielichhoferia elongata* moss shoot.

BELOW RIGHT | The so-called "copper moss" *Mielichhoferia elongata* growing on a steep shale slope.

RAINFORESTS

A general pattern in biogeography studies is the "latitudinal gradient of diversity," where the number of species increases on a gradient as one approaches the equator. Bryophytes, however, do not follow this rule, and tropical areas of the world are not richer in bryophyte species than temperate regions. In fact, lowland tropical rainforests, with their tall enveloping canopies and dark forest floors, are not especially rich in bryophytes at all. If altitude is added to the mix, though, and we consider the tropical forests that occur on mountainous areas near the equator, we find one of the richest and most diverse habitats for bryophytes on earth. In the montane rainforests, every branch is dripping with bryophytes growing in thick carpets or dangling in pendant wefts. In these hyper-diverse habitats, bryophytes play an important role in water cycling, since their extensive mass acts as a sponge, holding onto water to be released slowly and preventing the precious soils from being washed away by heavy rainfall.

Although the term "rainforest" immediately conjures an image of tropical forests, rainforests can also be found in the temperate regions of the northern and southern hemispheres, where rainfall is abundant and the climate is oceanic, meaning cool summers and mild winters. These habitats are incredibly rare and special places for bryophytes. The largest area of temperate rainforest in the world is in the Pacific coast region of Northwest America.

HARSH ENVIRONMENTS

The physiological adaptations of many bryophytes to enable their high tolerance of

the ice has recently been highlighted by scientists at the University of Alberta, who reported an outstanding discovery in 2013. The glaciers of the Canadian Arctic have been retreating, and this melting has been accelerating since 2004. On the remote Ellesmere Island, the glacier is gradually revealing more of the tundra floor as it retreats, and scientists were astounded to find clumps of mosses now exposed for the first time since they were entombed in ice. Radiocarbon dating confirmed that these plants were over 400 years old, and remarkably they were showing signs of new growth! The ultimate testament to the extraordinary survival qualities of bryophytes in some of the world's harshest environments.

desiccation means there are species able to thrive in some rather extreme conditions. Deserts may seem an unlikely home for bryophytes, but certain species have adapted to survive where water is in extremely limited supply. As early colonists of bare ground, they form an important component of soil crust communities that influence the absorption of water and prevent erosion, thus contributing to soil stability and providing micro-niches for soil invertebrates. By consolidating soil particles as they grow and by forming cushions and mats, they create an environment more hospitable to other plants.

Bryophytes are also found in the polar regions and may form the dominant vegetation in the tundra, beyond the point where trees can grow. Polar species are known to be tolerant of frost, but the extent to which they can survive under

AQUATIC ENVIRONMENTS

There are many bryophytes that are commonly found in very wet habitats, such as in bogs or on river boulders. These are mostly species tolerant of an aquatic lifestyle yet also found in drier conditions. Most need to be attached to a solid surface but there are some floating examples, such as the liverwort *Riccia fluitans*. Some species of moss can tolerate a coastal spray of seawater, but no bryophytes are truly marine.

Mosses have been found at surprising depths in lakes, forming thick moss mats on the lake sediments where the water clarity allows light to reach the depths. The sub-Antarctic lakes are well known for their aquatic mosses. The record for the deepest living moss found growing in these extraordinary lakes is *Bryum pseudotriquetrum*, which is recorded growing at depths of 265 ft (81 m) in Radok Lake, East Antarctica. This widespread species is more typically found in wet heaths or mountain streams. The adaptability of mosses to life underwater has meant they have been a popular choice for aquarist's fish tanks, and there is a trade in so-called aquatic moss plants for this purpose.

SPHAGNUM BOGS

Sphagnum is a unique and special plant that, in the right conditions and over long periods of time, can form extensive bogs or peatlands. This *Sphagnum*-dominated landscape covers an estimated 3 percent of the earth's surface, mostly concentrated in the temperate and boreal regions of the northern hemisphere, especially North America, Europe, and Russia.

The *Sphagnum* mosses have an inordinate capacity to absorb water, holding up to 20 times their own weight in water, and to some extent they can engineer their own optimal environment for growth. Most of the *Sphagnum* plant structure is

made up of large, empty dead cells with just a thin layer of narrow, living cells within the leaves. *Sphagnum* grows as a living wet blanket over the surface of the bog, holding huge amounts of water within its cells and keeping the surface layers of the bog completely sodden. Tree roots cannot grow in these waterlogged conditions, so potential encroachment of trees and shrubs is curtailed. *Sphagnum* plants are even capable of altering the pH of their bogs, making the water slightly more acidic through an electrochemical process carried out in the plant's cells that results in hydrogen ions being pumped into the surrounding water, increasing the acidity.

Below the surface, these waterlogged and slightly acidic conditions create an anaerobic environment, which means there is not enough oxygen available for the usual decomposing organisms to survive. Thus the old stems of the mosses and any other organic material subsumed within the bog decay very slowly. Scientists estimate that a depth of 3 ft (1 m) of peat can take up to 1,000 years to form since the organic matter is compressed beneath the bog and turns into deep layers of peat. The northern hemisphere peatlands store about three times as much organic carbon as the tropical rainforests, making *Sphagnum* and its peat a critical ally in the face of global warming.

LEFT | The dramatic waterfall of Fjaðrárgljúfur Canyon, Iceland, creates a habitat where high-humidity-dependant bryophytes thrive on the surrounding damp rocks.

BRYOPHYTES AND PEOPLE

Bryophytes have played an important role in people's lives for thousands of years, though in our modern-day world they often go unnoticed. Describing their vast diversity on earth is a major task for bryologists and one that increases in urgency as pressures on their natural habitats increase.

CULTURAL ASSOCIATIONS AND USES

Bryophytes were largely overlooked in European cultures until they started to gain more serious botanical attention in the eighteenth century. One exception is their application in the doctrine of signatures—a theory dating back to the fifteenth century that plants resembling certain features of the body could be used by herbalists to treat diseases of those body parts. Some of the larger mat-forming liverworts were observed with lobes reminiscent of the structure of the liver, hence the name "liver" wort (wort being an old English word for plant).

In other cultures around the world, bryophytes have a more long-standing appreciation. The Japanese national anthem "Kimigayo," based on poetry from the Heian period (794–1185), includes lyrics celebrating moss-covered rocks. Japanese moss gardens are often associated with Buddhist temples, and one of the most famous temples in Kyoto, "Kokedera," translates as "moss temple." These gardens are typically dominated by extensive mossy lawns usually incorporating the genera *Polytrichum*, *Dicranum*, and *Leucobryum*. Digging deeper into these cultural associations reveals that mosses have a role in customs around the world. In a tradition dating back to the twelfth century, the townspeople of Béjar in western Spain hold an annual procession of "Los Hombres de Musgo" in which locals dress from head to toe in suits of moss. This event commemorates the legend of when the town was recaptured from its occupiers by a band of local men able to surprise and overcome the occupying forces by creeping up camouflaged by mosses foraged from the local woods.

Sphagnum mosses, with their extraordinary capacity for absorbing liquids and their mild antiseptic qualities, have long been adopted as nature's hygiene products throughout history and are still used by some communities around the world today, who report fewer instances of diaper rash than with cloth-based alternatives. These useful properties of *Sphagnum* led to it being used as surgical dressing during war time, with significant organized harvesting undertaken in World War I. In modern times, the industrial-scale removal of peat from *Sphagnum* bogs, largely extracted for the horticultural trade but also used for fuel in some regions, has led to huge losses of this unique and precious habitat, leading to efforts to ban peat use for horticulture in some countries.

DESCRIBING THE BRYOPHYTE WORLD

The scientific names we use for different species of bryophytes are of relatively recent origins. The Italian botanist Pier Antonio Micheli (1679–1737) made some important first steps in the study of bryophytes and published accurate figures of several liverworts and the hornwort *Anthoceros*.

OPPOSITE | The Moss Garden at Saihō-ji Temple is dominated by moss genera such as *Polytrichum*, *Dicranum*, and *Leucobryum*.

The most significant early effort to describe the diversity of bryophytes dates back to German botanist Johann Dillenius (1684–1747), who moved to England in 1721. Dillenius's *Historia Muscorum*, published in 1741, described more than 661 species of "mosses" (under this broader concept he also included algae and lichens as well as other bryophytes), with illustrations and short descriptions. Dillenius took great care to observe and document these species with as much accuracy as possible, and this book served as the main guide to identify bryophytes for a long time.

The Swedish botanist Carl Linnaeus (1701–1778) included the bryophyte groups in his landmark publication *Species Plantarum* in 1753—a book that aimed to list every species of plant known at that time, and which also introduced the modern binomial naming system. Linnaeus's method of classification was very much based on the reproductive characters of flowers. He placed the bryophytes, along with other spore-producing organisms such as ferns and fungi, into his Cryptogamia section, which translates as "hidden marriage." This account was largely based on the work of Dillenius, whom Linnaeus visited in Oxford. *Species Plantarum* has been designated as the starting point for scientific names for the *Sphagnum* mosses, the hornworts, and the liverworts, meaning that any names that occurred in publications before this time are not valid.

Johannes Hedwig (1730–1799), a German botanist, was the first to interpret bryophytes more closely in line with our modern concepts. He used the microscope extensively and was an extremely accurate observer and illustrator. He was the first to correctly interpret the bryophyte life cycle and the function of the reproductive structures. He also noted the clear differences between mosses and liverworts for the first time. Hedwig published these findings, accompanied by his accurate and beautiful illustrations. His magnum opus publication *Species Muscorum Frondosorum* (1801) (published posthumously) is the designated starting point for the nomenclature of mosses other than *Sphagnum*.

The nineteenth century saw a significant increase in the number of bryophytes described from around the world. Unfortunately, differing species concepts to those we follow today meant that many names have since been reduced to synonymy, since the same species has been described multiple times from across its geographic range. Creating taxonomic order out of 250 years' worth of collective species descriptions is quite a task. Now, in the molecular era, analysis of DNA sequence data is providing an excellent tool to support this taxonomic work and has enabled greater insights into the past evolutionary relationships of these groups. But despite these technological advances, there remains a huge mountain to climb in terms of fully and accurately describing all bryophyte diversity. There are many large and complicated genera to resolve and new species to discover. Faced with the current biodiversity crisis, the question becomes one of how much of this bryophyte world can we still describe before pieces are lost forever. However, a clearer understanding of bryophyte diversity and distribution, both now and in the past, can help enormously in targeting conservation efforts.

OPPOSITE TOP | Detailed illustrations of two moss genera drawn by Johannes Hedwig for his publication *Species Muscorum Frondosorum* (1801).

OPPOSITE | Volunteer pickers of *Sphagnum* moss for Red Cross surgical dressings during World War I.

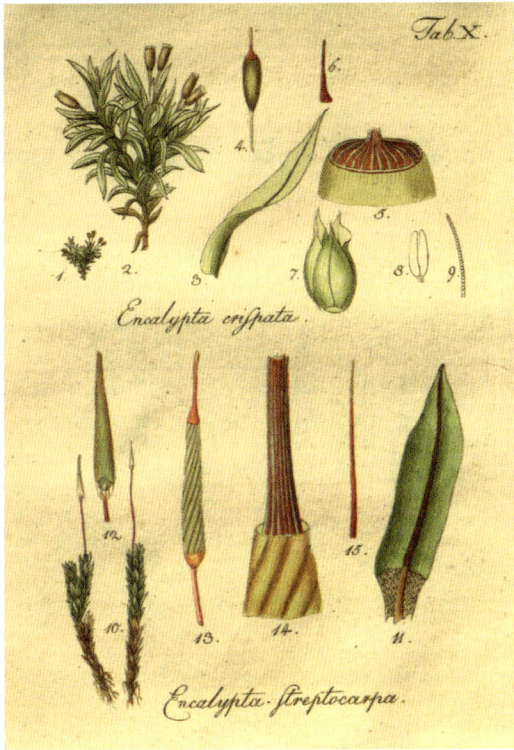

Tab. X.

Encalypta crispata.

Encalypta streptocarpa.

Tab. XLIII.

Bryum annotinum.

BRYOPHYTES IN THE ANTHROPOCENE

The last few hundred years of human activity have been so impactful on the planet that a new geological epoch has been defined; the Anthropocene. This reflects the extent to which Earth's physical, chemical, and biological systems have been altered. The last 60 years in particular have seen atmospheric carbon dioxide levels increase at an unprecedented rate, combined with the destruction and alteration of natural habitats.

ARE BRYOPHYTES AT RISK?

Bryophytes have survived major planetary changes before, but the challenge of the Anthropocene is the rate at which change is occurring, since vulnerable plants do not have the time to adapt to new conditions. Bryophytes are particularly at risk from global warming because many of them are dependent on a high-humidity microclimate and cool temperatures.

An assessment of the status of European bryophytes published in 2019 reviewed over 1,700 species and concluded that 22.5 percent of these were classified with some level of threat to their continued survival in Europe. This study benefited from Europe's long history of describing and mapping bryophyte diversity and its relatively high concentration of bryologists. In biologically diverse parts of the world where bryophytes face significant threats, such as loss of habitat in tropical forests, we do not have the detailed data required to produce these statistics. Threats to bryophytes can be complex, and it is sometimes hard to determine what the key driver for species threat might be among a number of possible drivers impacting a species.

THREATS TO BRYOPHYTES

The biggest immediate threat to bryophytes is the destruction, degradation, and fragmentation of their natural habitats. Landscapes have been physically modified by large-scale processes such as deforestation and urbanization. For example, the draining of wetlands around the world has had a huge impact on bog- and fen-dependent species. Pollution also remains a major threat to bryophytes. In the twentieth century, atmospheric sulfur dioxide pollution became a huge concern and significant improvements in reducing levels have since been made in many regions of the world. Now the most widespread pollution problem for bryophytes comes from nitrogen compounds, an over-abundance of which is released into the environment due to fertilizers from intensive agriculture and traffic fumes. This over-fertilization of the natural environment is leading to increased growth of nutrient-hungry green algae and cyanobacterial mats that can smother bryophytes even in remote upland areas. For bryophytes of nutrient-poor grassland habitats, nitrogen fuels the vigorous growth of weedy species that crowd out slower-growing plants of lower stature.

Climate change is impacting bryophytes across the world with warmer temperatures and increased droughts, in some regions also increasing the occurrence of wildfires. Alpine species, which require cooler conditions, have nowhere to go when the temperatures rise, and may be some of the first bryophyte casualties of a warming planet. *Sphagnum* mosses play an essential role in biochemical cycles by sequestering large quantities of carbon as peat in bogs. The extensive

peatlands of the boreal north store more than double the amount of carbon captured in tropical rainforests. Increasing global temperatures risks drying out and damaging these precious wetlands. The conversion of large areas of northern peatland from net sink to net source of carbon dioxide as a consequence of global warming is a major concern for the functioning of the planet.

A threat that people may be less aware of comes from the wild harvesting of bryophytes on a commercial scale. The destruction of *Sphagnum* bogs for peat extraction has been well documented. In some regions, large mats of bryophytes are also peeled from tree trunks and boulders of mossy forests for horticultural uses, festive decorations, and packing materials. This is not a regulated trade, and excessive removal of the plants can cause a lot of damage, with very slow recovery rates. Thanks to more recent awareness

ABOVE | This dramatic image of peat harvesting in Estonia illustrates the stark contrast between the diverse bogland habitat and the barren soil left after peat has been removed.

of conservation concerns, the harvesting of mosses is now illegal in some countries. Mosses can be sourced sustainably from nurseries that grow them or 'rescue' plants which, for example, may carpet an old garage roof about to be demolished. To ensure wild habitats are not being destroyed, it is always best to check the provenance of plants before purchase.

It is vital to understand threats to bryophytes so that we can then develop positive actions to counteract them and halt declines. This requires a detailed understanding of species requirements, their ecology, and the micro interactions within delicate ecosystems. For example, in Europe drastic declines in populations of the so-called

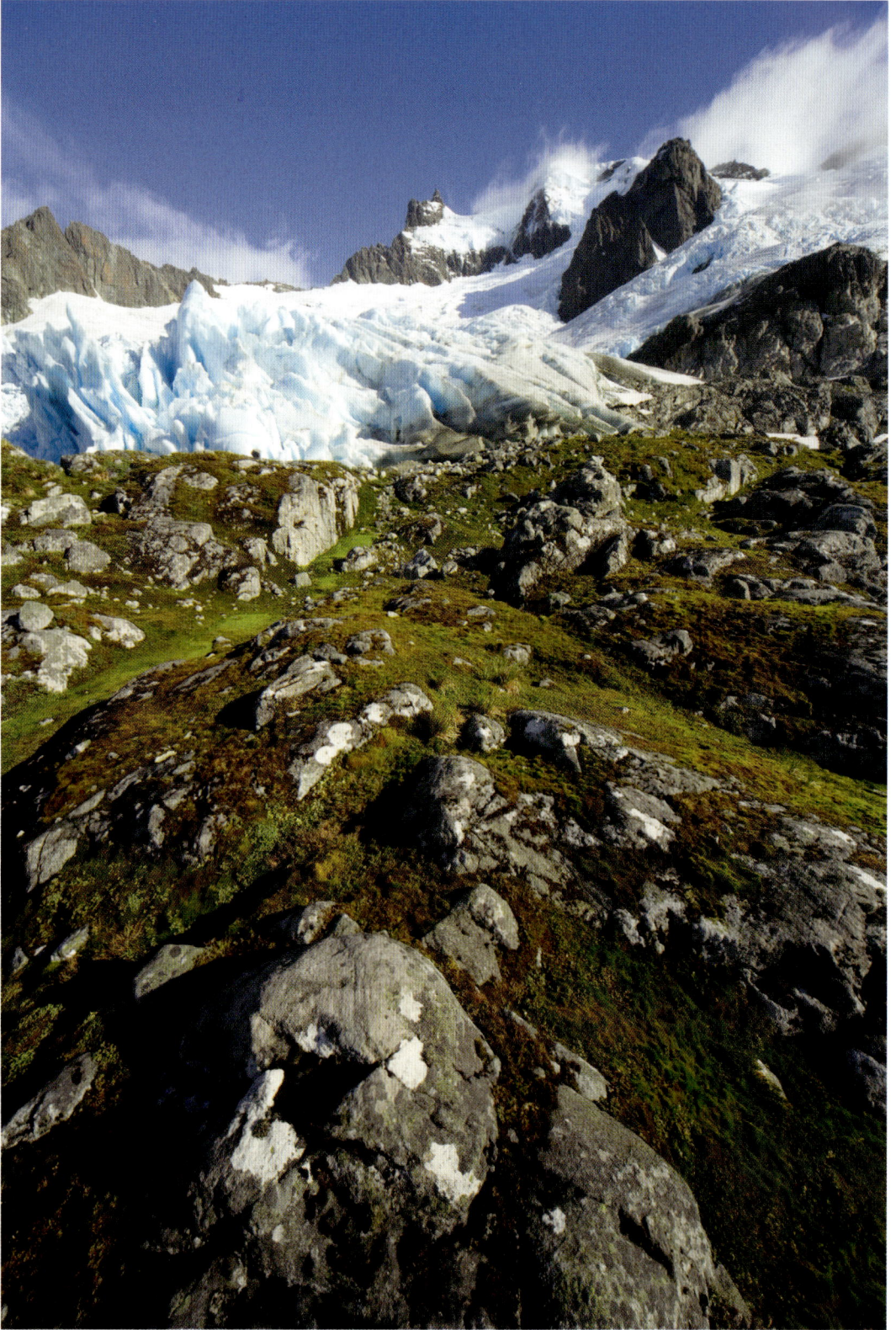

Bryophytes in the Anthropocene

"dung mosses" in the Splachnaceae family have been documented. Investigations have found that in some cases, the treatment of livestock with chemicals to treat parasite infections is the cause of the mosses' decline. These chemicals make the animal dung sterile, so affecting the invertebrates relying on dung for survival. The Splachnaceae mosses rely on flies feeding on the dung to disperse their spores, and the removal of their vector has limited their ability to reproduce.

HOW TO PROTECT BRYOPHYTES FOR THE FUTURE

Despite the concerning headline news, there are also success stories that show how conservation action can work. There are examples where damaged habitats have been restored, though the timescale for recovery can be slow. Blocking drains has allowed degraded peat bogs to stabilize and begin to recover, aided at times by the transplanting of propagated *Sphagnum* plants. Excluding grazing animals from sites that have been degraded by over-grazing has in places shown remarkable results, allowing a landscape to regrow.

There are still significant gaps in our knowledge of diverse regions of the world that lack up-to-date checklists of the bryophytes found there. Data on the detailed distribution and occurrence of species is often very limited. This is fundamental data needed to make assessments on the conservation status of species. Dedicated community scientists or amateur botanists can make a significant impact here by learning how to identify local bryophytes and recording their occurrence for regional mapping schemes. Gaining a more detailed picture of past and present species distribution will help to identify bryophytes in need of conservation action and could support the protection of important bryophyte sites with valuable data. Creating protected areas that include bryophyte-rich habitats and implementing a management strategy for these sites will go a long way toward safeguarding bryophyte diversity into the future.

OPPOSITE | This glacier is receding over bryophyte-covered tundra slopes on the sub-Antarctic South Georgia Island, a harsh habitat dominated by bryophytes and lichens.

BELOW | The so-called "dung mosses," such as this *Splachnum rubrum* moss, are reliant on insects visiting their brightly colored capsules and dispersing the spores.

BRYOPHYTE DIVERSITY

On the preceding pages we have seen how diverse the bryophytes are, but how is this wealth of diversity communicated? This is the job of the taxonomist. Taxonomy is the science of naming, describing, and classifying life on earth. We still follow the classification system introduced by Linnaeus in the eighteenth century, which sorts groups of species into smaller and more similar groups. The major categories of this hierarchical classification system, from largest to smallest, are:
Kingdom > Phylum > Class > Order > Family > Genus > Species

The classification of bryophytes continues to be refined as new findings piece together how these plants are related to each other. In this book we will see that we are often unsure about the number of species in many bryophyte genera, and this may seem like an odd thing for us not to know. Over the past 200 years, botanists who study bryophytes—bryologists—have been formally documenting and describing species, with a large proportion of these names described in the nineteenth century. Unsurprisingly, many species have inadvertently been described more than once, and an important job of the taxonomist is to continue editing this manuscript of nature so that unneeded synonymous names are identified and carefully removed. Exploration of the bryological world is still not complete, with new species continuing to be discovered and DNA sequence data increasingly revealing cryptic species that had not previously been recognized. Accurately documenting the global diversity of bryophytes and where they occur around the world remains a huge task, which bryologists are still working to achieve.

Kingdom

All bryophytes belong in the kingdom Plantae. They share a common ancestor with other members of the Plantae kingdom such as the gymnosperms and angiosperms. Their common ancestry can be seen at a cellular level, for example in the origins of their photosynthetic chloroplasts.

Phylum

In bryophytes, this level of classification separates out the three different groups of mosses, liverworts, and hornworts. The origins of these three groups of plants are discussed on pages 10–11.

Class

Bryophytes that belong to the same class share some fundamental features. For example, the Polytrichopsida mosses are defined by the unique structure of their peristome teeth, and they are also confirmed to be a distinct evolutionary lineage.

Order

Bryophytes that are classified in the same order will share some important features and a common ancestor. Some orders are huge and some very small, in some cases with just one species as the only living representative of a long evolutionary lineage.

Family

At the level of family, the plants should usually share a number of common features that would enable them to be recognized as belonging to that

OPPOSITE | *Zygodon rupestris*, a common European species in the Orthotrichales that can be found typically growing on tree trunks.

family, though they could still show a lot of diversity between genera.

Genus

Generally speaking, bryophytes should be identifiable to genus level with a few morphological characters. Genera should also reflect a grouping with a shared ancestry, and this has led to something of a shake-up in larger genera that have been found to be derived from more than one ancestral group.

Species

At the most basic level, a species represents the groupings a person would recognize when separating out different types of plants based on their shared characters. It is an extremely important level of classification, as it is a concept most people have some familiarity with and is used extensively in conservation metrics and targets.

BELOW | *Tetraphis pellucida* is a common moss of temperate northern hemisphere.

Bryophyte Diversity

THE FULL LINNEAN CLASSIFICATION
OF THE MOSS *TETRAPHIS PELLUCIDA*

Kingdom

Plantae

Phylum

Anthocerotophyta   Marchantiophyta   Bryophyta

Class

Takakiopsida   Sphagnopsida   Andreaeopsida   Andreaeobryopsida   Polytrichopsida   Bryopsida

Order

Oedipodiales   Tetraphidales   Polytrichales

Family

Tetraphidaceae

Genus

Tetraphis   *Tetrodontium*

Species

Tetraphis pellucida   *Tetraphis geniculata*

SELECTED GENERA

At the time of writing, the diversity of bryophytes is classified into 73 orders, 1,344 genera, and over 15,000 species. The classification used here is based on the latest research from the international community of bryologists and represents a phenomenal amount of research culminating over many decades.

Every order of bryophytes across the hornworts, liverworts, and mosses is represented in this section by at least one genus and the more diverse orders by two or three genera. An exception is made for the Hypnales, where six genus profiles have been included since this order of mosses has the most genera of any bryophyte order.

For those orders that include only one genus, the decision was easy! Where genera had to be selected from a range of potential candidates, the following criteria were considered. Some genera are so charismatic that they simply had to be included. For example, *Schistostega*—a moss that glows in the dark, no less—was essential to feature in the Dicranales along with other examples of this order. An attempt was made to include the most species-diverse and globally widespread genera—such as *Bazzania*, *Herbertus*, and *Plagiochila*—within the liverwort order of the Lepidoziales. Particularly striking and beautiful genera were included to illustrate the orders where possible—such as *Dawsonia*, the world's tallest moss. The 100 genera chosen here aim to illustrate the rich structural and evolutionary diversity of bryophytes from around the world.

Notes on the profiles

Each genus profile includes an information panel highlighting key features as well as a small distribution map.

The estimated number of species described for each genus is based on a wonderful online resource provided by the Missouri Botanical Garden based on their immense nomenclatural database Tropicos. It is worth noting, of course, that the number of species names accepted across the scientific literature may not tally with the actual number of "real" species. These figures are constantly changing as new findings are published.

BELOW LEFT | The liverwort *Bazzania trilobata* has a dichotomously branched or Y-shaped leaf pattern.

BELOW RIGHT | The moss *Ptilium crista-castrensis* has pinnate or feather-like branching.

COMMON MOSS LEAF SHAPES

Linear—long and narrow with pointed tip	Lanceolate—wider at the base and tapering to a point	Ovate—egg-shaped with a pointed tip	Lingulate—tongue-shaped with parallel sides	Spathulate—wider above the middle than below

The maps serve to include the global range of the plants and are based on the available scientific literature. One of the challenges here is that a genus may be described in print as occurring across Africa, Southeast Asia, and Australia, but when it comes to imprinting this description on a map, some important details are missing! While every effort was made to ensure accuracy at this scale, the maps represent the level of information available at the time of writing.

TAXONOMIC TERMINOLOGY

All natural-history enthusiasts employ a certain amount of jargon for their areas of special interest and bryologists are no different! Taxonomic jargon is kept to a minimum in this book but some scientific terms are necessary. Some of the terms that occur repeatedly through these pages are explained in more detail below.

Leaf shape

There are a number of commonly used terms to describe the shape of bryophyte leaves, and the most frequently used of these are illustrated above.

Branching patterns

Bryophytes can show a variety of branching patterns (see opposite), which are informative taxonomic characters.

Sexual condition

Whether plants produce male or female sexual organs—or both—is an important taxonomic characteristic. If the male and female sex organs are produced on different plants, they are called dioicous; if they are produced on the same plant, they are called monoicous. Monoicous plants can further be divided into those that produce the sex organs on different branches (autoicous) and those that produce the sex organs together in the same cluster (synoicous).

Bryophytes that grow on other plants

Many species of bryophytes grow directly on other plants, and these are called epiphytes. They are using the trees and shrubs simply as a convenient structure to perch themselves on. Those species that grow specifically on leaves are called epiphyllous.

Parasitic and symbiotic relationships

Many liverworts have symbiotic associations with fungi, which means both partners gain something from the relationship. Only the liverwort *Aneura mirabilis* is known to have a parasitic relationship, in this case with a fungus, from where it sources its carbon.

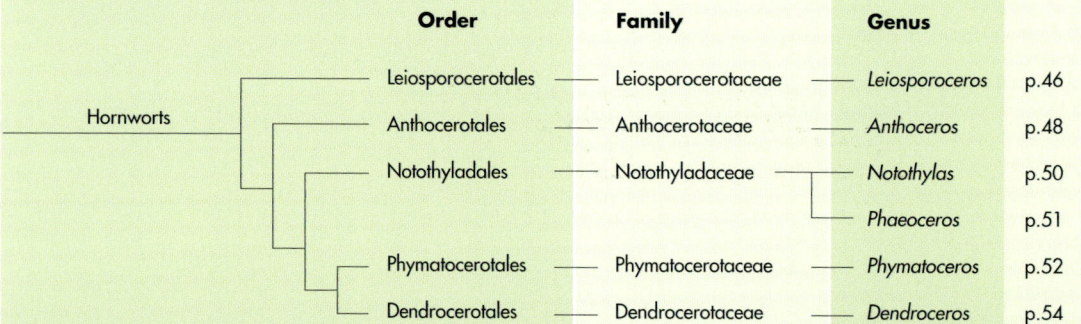

ANTHOCEROTOPHYTA: THE HORNWORTS

The hornworts have a number of unique aspects to their biology. Most have a symbiotic association with cyanobacteria; their large solitary chloroplasts have similarities to those of the algae, and the narrow, elongate "horns" of the spore-producing structures are immediately recognizable. In fact, they are so distinct from any other land plant lineage and yet so sparse in distinct taxa compared to other groups that some scientists suggest they may have been more diverse in the past.

Hornworts are the least taxonomically diverse of the three bryophyte groups, with an estimated 200 to 250 species classified within 5 orders and 14 genera. The first two decades of the twenty-first century have been an exciting time for hornwort research, with major new insights into their structures, evolutionary relationships, and classification.

The Leiosporocerotales are considered particularly distinct and a sister group to all other hornworts. The characters used to define the different orders of hornworts mainly rely on a combination of DNA sequence data to reveal past evolutionary relationships and on microscopic anatomical features such as details of the spores and features of the chloroplasts.

LEIOSPOROCEROS

The species *Leiosporoceros dussii* is considered so unique among hornworts that it warrants its own order and is considered a sister group to all other hornworts because it is both morphologically and genetically distinct.

The plants form elongate rosettes and the projecting cells on the surface give the plant a velvety feel. Most striking are the long blue lines drawn

BELOW | A spectacular colony of *Leiosporoceros* showing the elongated branches of the thallus and abundant sporophytes.

DISTRIBUTION
Mexico, Central America (Costa Rica, Panama), northern Andes, Lesser Antilles

ETYMOLOGY
Greek *leios* = "smooth" + *kéras* = "horn," referring to the smooth unornamented spores

NUMBER OF SPECIES
1 accepted species

APPEARANCE
Robust, fleshy thallus forming rosette, branches ¾ in (2 cm) in diameter. Cells 1–2 chloroplasts, no pyrenoid. Mature sporophytes 1½ in (4 cm) long. Pseudoelaters long, thick-walled. Antheridia up to 70 per chamber. Cyanobacterial *Nostoc* colonies visible as blue strands

HABITAT
On rocks or sandy soils, usually near water bodies

along the thallus. These are *Nostoc* colonies oriented longitudinally inside the thallus in mucilage-filled canals. This linear arrangement of the colonies running parallel to the main axis of the thallus is not only unique among the hornworts but is unknown in any other land plant. These colonies initiate when *Nostoc* enters the thallus and establishes behind the cell at the growing tip of the thallus. As the thallus then elongates, so does the *Nostoc* colony within an advancing canal of mucilage.

Leiosporoceros also has distinctive spores that are nearly smooth, transparent, and slightly elongated. When spores are being formed they are initially attached to each other in groups of four. When these are formed in a tetrahedral arrangement they leave a trilete (Y-shaped) scar on each spore from where they

were initially joined with three other spores. If they originated in a tetragonal formation then each of the four initial spores would have been in contact with only two of their neighbors, and the resulting scar is called monolete (linear-shaped). Most hornworts have spores with a trilete mark except for *Leiosporoceros*, which have the monolete mark, demonstrating a clear divergence in early development processes.

ANTHOCEROS

The Anthocerotales includes just the two genera: the widespread *Anthoceros* and the tropical *Folioceros*. *Anthoceros* was the first hornwort genus ever to have been described and features in Linnaeus's landmark publication *Species Plantarum* in 1753. It is also the most species-rich hornwort genus, with over 60 species recognized.

Although some well-known species of *Anthoceros* can be distinguished by features seen with a hand lens, some species of *Anthoceros* from tropical regions can only be identified by examining minute characteristics of the dark-brown spores. Other features needed to identify this order include the details of the male reproductive structures, where there are numerous antheridia in each chamber, contrasting with most other hornwort orders that only have one or two antheridia in each chamber.

Anthoceros agrestis is increasingly being used as the model system for the study of hornwort biology, plant and cyanobacteria interactions, biochemistry, and detailed genomic information. It is a plant that copes very well as a laboratory rat due to its small size and easy propagation. Its genome has been sequenced and is providing insights into the innovations for early land plant evolution. There is also future potential to explore genetic engineering by transferring useful hornwort traits into crop plants to increase yields, such as efficient mechanisms for photosynthesis using pyrenoids or the symbiosis with cyanobacteria. These detailed studies into the mechanics of hornwort biology have the potential to dramatically impact various fields and support food security in the face of increasing pressures on agriculture.

LEFT | Colonies of the cyanobacterium *Nostoc* are visible as dark spots on the surface of the hornwort thallus.

DISTRIBUTION
Globally widespread, temperate, tropical

ETYMOLOGY
Greek *ánthos* = "flower" + *kéras* = "horn," referring to the horn-shaped sporophytes

NUMBER OF SPECIES
67 accepted species

APPEARANCE
Dark-green rosettes, frilly margins, no midrib. Sporophytes $3/8$–2 in (1–5 cm) tall. Dark-brown spores, trilete mark, bumps on surface. Cells 1–4 chloroplasts with/without pyrenoid. Antheridia 4–45 per chamber. Thallus with mucilage-containing cavities. *Nostoc* scattered across thallus surface visible as black dots

HABITAT
Pioneer of open environments on disturbed ground such as arable fields, banks, and path sides

LEFT | *Anthoceros punctatus* is a widespread species that typically forms small rosettes with frilly margins.

BELOW | *Anthoceros fusiformis* can be distinguished by the upright flaps of thallus tissue, as can be seen in the lower part of this image.

NOTOTHYLAS

Ntotothyladales is an order particularly notable for the heterogeneity of its four genera. Within the order, *Notothylas* is distinguished by a number of features, but most apparent is the length of the sporophytes, which are the shortest of all the hornworts.

The remaining features are very variable even within the genus. A particularly unusual variation is the spore color, which is usually consistent within hornwort genera but here can vary from yellow to dark brown or black. In an attempt to bring some order to this rather chaotic grouping, taxonomists have previously proposed various subsections to divide up the genus, though analysis of DNA sequence data does not support these divisions. *Notothylas* is defined as monophyletic, meaning the taxonomic grouping includes a single common ancestor plus all of its descendants, albeit one with a wide range of defining characteristics.

Despite its lack of conformity, *Notothylas* is an easy genus to recognize due to its short, stumpy sporophyte "horns" lying almost flat on the thallus surface. It is widely distributed in tropical to temperate regions, with its highest diversity across the Indian subcontinent.

LEFT | Rosettes of *Notothylas orbicularis* with the tube-like involucres that protect the developing sporophyte laying flat against the thallus.

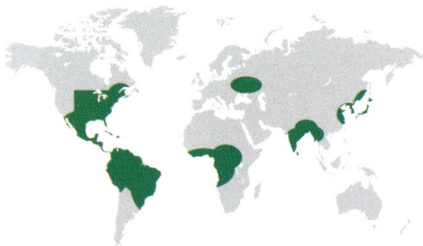

DISTRIBUTION
A pantropical and northern-temperate genus

ETYMOLOGY
Greek *nôto* = "dorsal" + *thylakos* = "pouch," referring to position of sporophyte on dorsal surface of thallus and protective structures enclosing the developing sporophyte

NUMBER OF SPECIES
26 accepted species

APPEARANCE
Small rosettes ³⁄₁₆–1¼ in (0.5–3 cm) in diameter. Cells contain 1–3 chloroplasts. Sporophytes enclosed within the involucre. Spores with trilete (Y-shaped) mark, pseudoelaters without spiral band. Antheridia 2–6 per chamber. Globose *Nostoc* colonies across thallus, visible as black dots

HABITAT
Disturbed soil in open habitats

PHAEOCEROS

Demonstrating just how variable the Notothyladales really are, the genus *Phaeoceros* looks strikingly different from *Notothylas*. These larger plants are ¾–1¼ in (2–3 cm) in diameter and can be rather robust and fleshy. The sporophytes grow long and upright, sometimes reaching heights of 3½ in (9 cm), though they are usually more modest.

Despite being classified in a different order, *Phaeoceros* can look very similar to *Anthoceros* in the Anthocerotales and the two genera can be easily confused when they grow together in similar habitats. *Phaeoceros* can be distinguished in the field by its yellow spores in contrast to the blackish spores of *Anthoceros*. On a more microscopic level *Phaeoceros* lacks the numerous large mucilage-containing cavities characteristic of *Anthoceros*.

Phaeoceros has a worldwide distribution, with highest species diversity in the tropics, where examining details of the spores may be needed to identify species.

ABOVE | *Phaeoceros minutus* on bare soil of the Western Cape, South Africa.

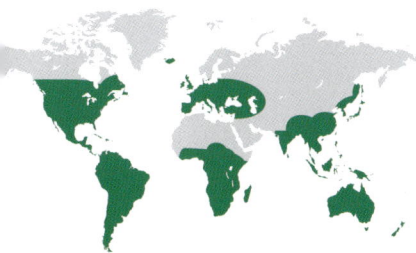

DISTRIBUTION
Worldwide

ETYMOLOGY
Greek *phaeo* = "dark colored" + *kéras* = "horn," although genus described with yellow spores

NUMBER OF SPECIES
35 accepted species

APPEARANCE
Dark green with thallus forming rosettes ¾–1¼ in (2–3 cm) diameter. Cells contain 1–2 chloroplasts. Sporophytes 1–4 up to 3½ in (9 cm) tall. Spores with spinose to bumpy ornamentation. Antheridia 2–4 per chamber

HABITAT
On soil in open environments

PHYMATOCEROS

*P*hymatoceros was only established in 2005 when subtle yet significant differences in a species previously described as *Anthoceros bulbiculosus* were considered to be sufficiently different from all other hornwort genera to justify creating a new genus. Since then, two additional species have been assigned to this genus, with one species, *Phymatoceros binsarensis*, described in 2023 based on plants discovered in India, extending the distribution of this genus from the Americas, Europe, and Africa eastward into Asia. Further analysis of hornwort classification has found this small genus to be so different from other hornworts that it is placed in its own order, the Phymatocerotales. Two of these species had previously been placed in *Phaeoceros*.

Rather than forming the typical rosette of most hornworts, *Phymatoceros* grows in a more linear fashion, with a narrow parallel-sided branching thallus. Another striking feature is the prolific long, stalked tubers in the upper region of the thallus. There are one to two chloroplasts per cell, but in contrast to most other

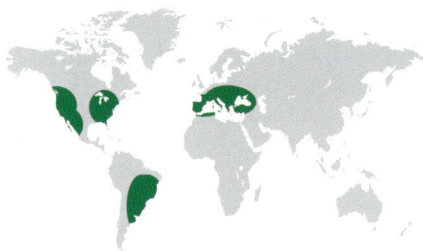

DISTRIBUTION
Widespread in Mediterranean Europe and Africa, India, and North and South America

ETYMOLOGY
Greek *phȳma* = "tumor" or "growth" + *kéras* = "horn," in reference to the prolific number of tubers produced

NUMBER OF SPECIES
3 accepted species

APPEARANCE
Narrow linear thalli up to ⁹⁄₁₆ in (14 mm) long. Cells 1–2 chloroplasts with or without a pyrenoid, depending on species. Spores brown-black, bumps/dimples distinguish species. Pseudoelaters short, thin-walled. Antheridia 1 or 2–5 per chamber. Globose *Nostoc* communities on thallus

HABITAT
A colonizer of bare soil in open environments

hornworts, not all species in *Phymatoceros* possess a pyrenoid. These species are dioicous, meaning that male and female reproductive organs are produced on separate plants. The female plants are slightly smaller than the males and produce a short robust sporophyte. Although united by their molecular data and some morphological similarities, the three species placed in this order are morphologically quite divergent from each other, and summarizing diagnostic features for the genus is challenging.

DENDROCEROS

The Dendrocerotales differ from the other hornwort orders in their choice of habitat as they have specialized to grow on living plants as epiphytes rather than as colonizers of bare soil. They may even grow on the surface of leaves in areas with high humidity. The Dendrocerotales are predominately tropical, and *Dendroceros* is one of four genera recognized in the order. *Dendroceros* is easily recognized by the crisped thallus that is differentiated into a central thickened midrib and lateral wings that may be perforated with holes.

BELOW | Under the microscope these green *Dendroceros* spores can be seen to be multicellular, intermixed with the long, spiraled pseudoelaters.

The spores, which are so heavily relied upon in hornwort taxonomy, are particularly interesting in *Dendroceros* since they exhibit endosporic germination, which means the spore cell starts dividing within the cell wall before a protonemal filament emerges, resulting in multicellular spores. For a plant that grows in rather precarious habitats prone to desiccation, such as on the surface of a tropical tree leaf, it is easy to see the advantage of getting an early start and reaching

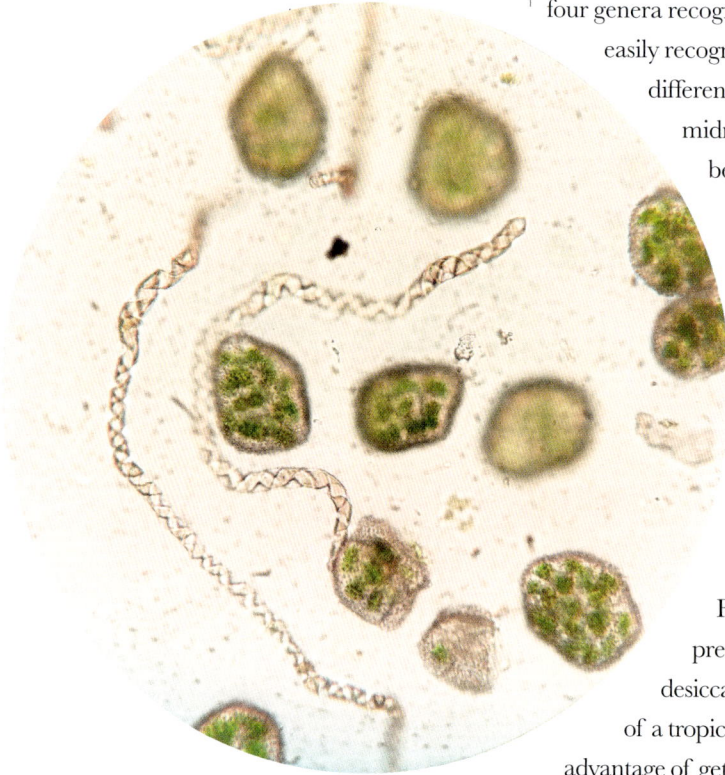

DISTRIBUTION
Widespread across tropical and extending into subtemperate regions

ETYMOLOGY
Latin *dendro* = "trees" + *kéras* = "horn," referring to the epiphytic habit of these plants

NUMBER OF SPECIES
Around 39 accepted species

APPEARANCE
Yellow-green. Solid thallus $1/16$–$3/16$ in (2–5 mm) wide, midrib, lateral wings. Cells 1 large chloroplast. Mature sporophytes $3/8$–2 in (1–5 cm) tall. Pseudoelaters narrow, spiraled. Spores colorless, yellow, pale green. Antheridia 1–2 per chamber. *Nostoc* on thallus appear as black dots

HABITAT
Twigs, branches, living leaves in humid temperate/tropical forests

maturity to reproduce as soon as possible. The spores of the Dendrocerotales are colorless or yellow-green, which is typical of the tropical species associated with this order, where the spores are short lived. Darker spores typical of other hornwort orders have thicker walls and are filled with lipids, which means they are more desiccation tolerant and longer lived.

PHYLOGENETIC TREE OF LIVERWORT
ORDERS AND FEATURED GENERA

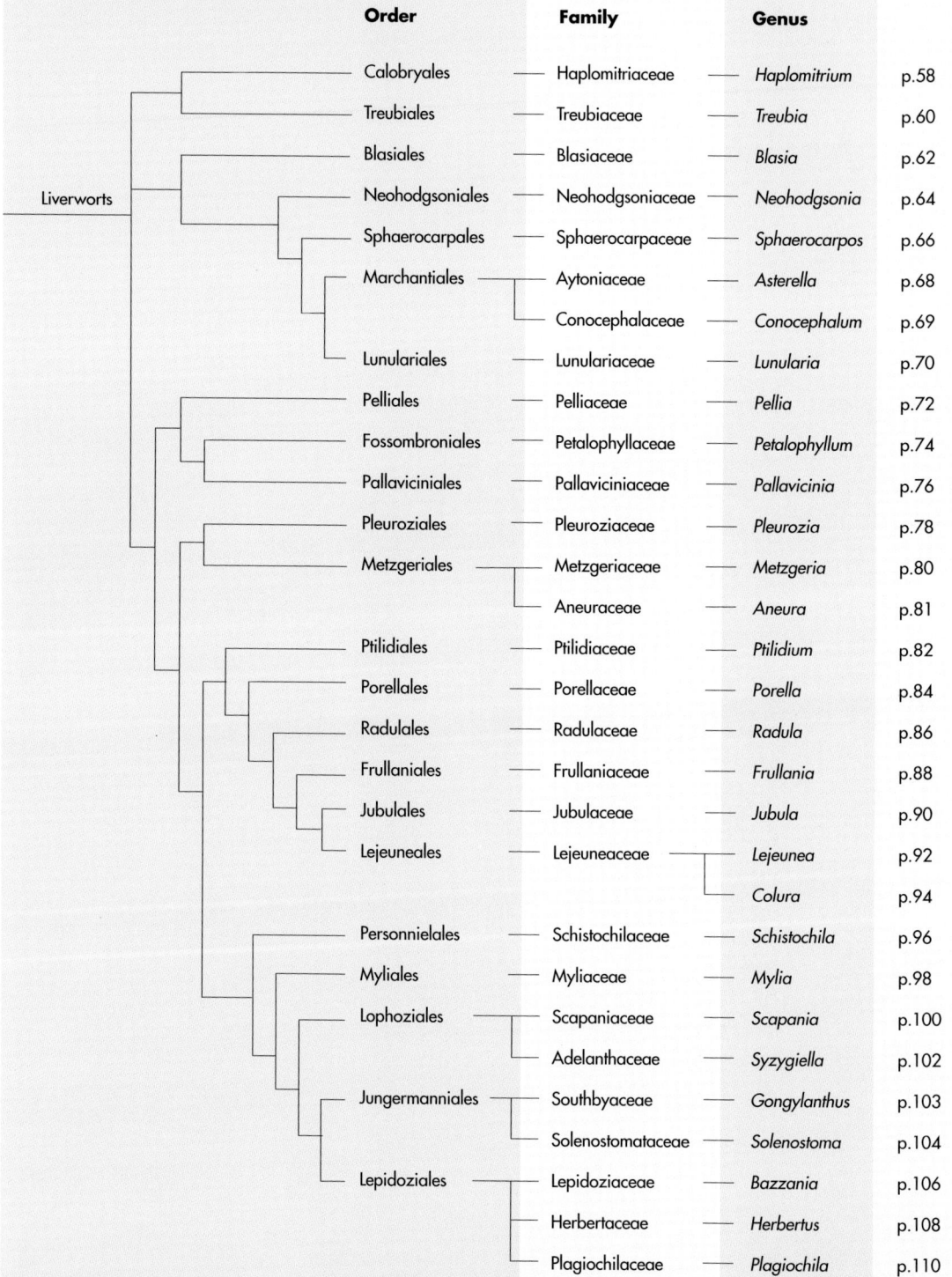

Order	Family	Genus	
Calobryales	Haplomitriaceae	*Haplomitrium*	p.58
Treubiales	Treubiaceae	*Treubia*	p.60
Blasiales	Blasiaceae	*Blasia*	p.62
Neohodgsoniales	Neohodgsoniaceae	*Neohodgsonia*	p.64
Sphaerocarpales	Sphaerocarpaceae	*Sphaerocarpos*	p.66
Marchantiales	Aytoniaceae	*Asterella*	p.68
	Conocephalaceae	*Conocephalum*	p.69
Lunulariales	Lunulariaceae	*Lunularia*	p.70
Pelliales	Pelliaceae	*Pellia*	p.72
Fossombroniales	Petalophyllaceae	*Petalophyllum*	p.74
Pallaviciniales	Pallaviciniaceae	*Pallavicinia*	p.76
Pleuroziales	Pleuroziaceae	*Pleurozia*	p.78
Metzgeriales	Metzgeriaceae	*Metzgeria*	p.80
	Aneuraceae	*Aneura*	p.81
Ptilidiales	Ptilidiaceae	*Ptilidium*	p.82
Porellales	Porellaceae	*Porella*	p.84
Radulales	Radulaceae	*Radula*	p.86
Frullaniales	Frullaniaceae	*Frullania*	p.88
Jubulales	Jubulaceae	*Jubula*	p.90
Lejeuneales	Lejeuneaceae	*Lejeunea*	p.92
		Colura	p.94
Personnielales	Schistochilaceae	*Schistochila*	p.96
Myliales	Myliaceae	*Mylia*	p.98
Lophoziales	Scapaniaceae	*Scapania*	p.100
	Adelanthaceae	*Syzygiella*	p.102
Jungermanniales	Southbyaceae	*Gongylanthus*	p.103
	Solenostomataceae	*Solenostoma*	p.104
Lepidoziales	Lepidoziaceae	*Bazzania*	p.106
	Herbertaceae	*Herbertus*	p.108
	Plagiochilaceae	*Plagiochila*	p.110

Liverworts

MARCHANTIOPHYTA: THE LIVERWORTS

The liverworts are morphologically an extremely diverse group, with a vegetative body plan varying from green thalloid plates to leafy shoots and a variety of complex leaf structures. The shared ancestry of the Marchantiophyta only really becomes apparent when the plants produce sporophytes and the uniform features of the setae and capsules become apparent.

There are around 7,000 known species of liverworts, which are recognized in 23 orders and 363 genera. The evolutionary relationships within the liverworts have been a puzzle to decipher. Some lineages have lost features and have become more structurally simple than their ancestors, and leaf-like structures have actually evolved more than once within the liverwort dynasty. But DNA sequencing has helped reach a more stable classification reflective of the relationships and origins of these plants.

Although the liverworts have ancient origins, some of the most species-diverse orders—such as the Lejeuneales and Lepidoziales—evolved more recently in parallel with the rise of the angiosperms and the tropical forests. These groups were able to diversify rapidly into these new habitats and are a significant part of forest epiphyte communities to this day.

HAPLOMITRIUM

Calobryales is a very ancient order that diverged early in the evolutionary history of the liverworts. *Haplomitrium* is the only living representative of this order, giving it a special significance in the study of early land plants.

The green leaves are positioned in three rows up the stems, giving a spiraled appearance. These small leafy shoots look very simple, but that overlooks some rather advanced adaptations at a cellular level. Some species have a central strand in the stem made up of long narrow cells that act as an internal water-conducting system. The lower part of the plant consists of a unique thick colorless rhizome, the branches of which bury themselves in the substrate. Symbiotic associations with fungi are common across the liverworts,

BELOW | A single plant of *Haplomitrium hookeri*, showing the pale rhizome from which the leafy shoots arise.

OPPOSITE | Typically found on damp ground, such as lake margins, these tiny green plants require careful searching on hands and knees to find.

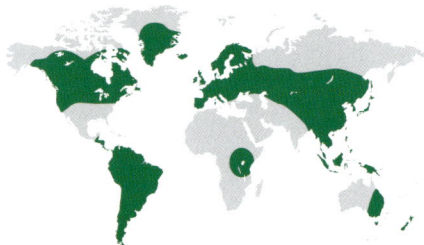

DISTRIBUTION
North and South America, Europe, Asia, Africa, Australia, Pacific Islands

ETYMOLOGY
Greek *haplóos* = "simple" + *mitrium* = "little cap," referring to the calyptra that functions as a protective sleeve for the sporophyte

NUMBER OF SPECIES
7 accepted species

APPEARANCE
Up to ¾ in (2 cm) tall, 2⅜ in (6 cm) for some species. Leafy shoots, rounded leaves, pale root-like rhizome. Capsules cylindrical. Antheridia at shoot tips, ball-like, often orange, in leafy splash cups

HABITAT
Mostly soil or gravel in humid habitats but also wood or rocks

and these underground branches are colonized by fungi. These are being studied to provide insights into how the earliest land plants may have adapted to terrestrial habitats with the help of fungal partners.

The sporophyte is also unique and surrounded by a fleshy calyptra while developing. The capsules are cylindrical, which is also striking as almost all other liverwort capsules are round. The setae are unusually thick and can extend up to 1 ¼ in (3 cm).

The greatest species diversity for the genus is in the southern hemisphere. Although only seven species are accepted, these show high levels of genetic diversity, and it is likely more species will be recognized with future study.

TREUBIA

The Treubiales represent an ancient group that, along with the Calobryales, diverged very early on in the evolution of the liverworts. The liverworts broadly fit into two structural groups: the thalloid liverworts, which form flat green mats without stems or leaves, and the leafy liverworts,

BELOW | A close-up view of the lobule of a *Treubia lacunosa* plant; the yellow dots are the large oil bodies.

DISTRIBUTION
Only found in the southern hemisphere—Chile, Southeast Asia, Australasia, and South Pacific islands

ETYMOLOGY
After Melchior Treub (1851–1910), a Dutch botanist

NUMBER OF SPECIES
7 accepted species

APPEARANCE
Thallus fleshy, brittle, bright green with yellow dots of oil body cells on the surface, branches up to 4 in (10 cm) long and ⅝ in (1.5 cm) wide with 2 rows of leafy lobes along the axis. Leaf cells dimorphic. Dioicous. Sporophytes arising from shoot tips. Asexual reproduction by gemmae

which have stems and leaf-like structures. The Treubiales is a particularly fascinating order from an evolutionary perspective because the structural complexity of these plants appears intermediate between these leafy and thallose forms. The result is a thalloid plant's innovation for a version of leaves!

Treubia has a unique leafy growth form for what is essentially a thalloid body plan. The thallus has a thick central axis with two rows of leaf-like lobes and smaller dorsal lobes along the upper surface. The leaves have two kinds of cells, some containing a large oil body scattered among other cells containing chloroplasts. The upper surface of the thallus is covered in a layer of mucilage produced from specialist mucilage-secreting cells. Sporophytes are positioned at the end of the thallus branches and have an unusually large seta.

There are two extant genera within the Treubiales: *Treubia* and *Apotreubia*. *Treubia* is a remarkable genus in terms of its unique morphology and evolutionary significance. It is distributed across the southern hemisphere, reflecting a time when the southern continents were united in the ancient supercontinent of Gondwana.

ABOVE | The structure of *Treubia tasmanica* illustrates how these plants are intermediate between leafy and thalloid liverwort forms.

HABITAT
Growing over soil or rotting logs in humid sites, tropical forests

BLASIA

The Blasiales have a series of features that make them unique among the liverworts. Symbiotic relationships with cyanobacteria are common among the hornworts but in liverworts are restricted to the Blasiales, where the scattered dark spots of *Nostoc* colonies can be seen on the surface following a single line on each side of the thallus.

Blasia produces two distinct kinds of gemmae that play different roles in its vegetative reproduction. The green spiky gemmae balls produced on the upper thallus surface are short lived and develop in the summer months when they disperse and grow into new plants. These gemmae plan ahead and carry within them a tiny piece of the *Nostoc* symbiont to pastures new. The smooth, pale-yellow gemmae are produced in long, narrow, flask-shaped receptacles and contain nutrient reserves allowing them to overwinter in a dormant state and then start to grow in the spring. These are initially lacking a *Nostoc* symbiont, which needs to be acquired later. Sometimes both types of gemmae are produced on the same plant. All the sporophyte-bearing gametophyte thalli die off in the fall and the green sporophyte stays protected among the dead thallus throughout the winter, maturing and producing its spores in the following spring.

Blasia shares some features with the more complex thalloid liverworts. Its short, linear branches with lobed and ruffled margins have two rows of scales on the underside and are similar to those in Marchantiales. However, the thallus of *Blasia* is not internally differentiated with air chambers and pores.

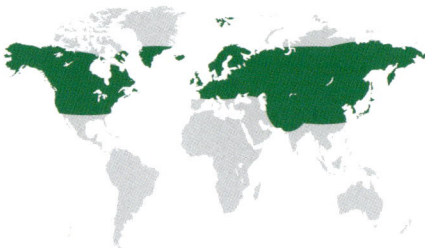

DISTRIBUTION
Circumpolar boreo-temperate—Europe, Russia, Himalaya, China, Japan, North America

ETYMOLOGY
After Blasius Biagi (c.1670–1735), a Benedictine monk and botanist

NUMBER OF SPECIES
Only 1 species

APPEARANCE
Thallus branches to 1 in (2.5 cm) long with midrib and lobed margins. Dioicous. Antheridia arranged in two rows embedded in thallus surface. Sporophytes at tips of thallus lobes, large seta, capsule with elaters. Two kinds of gemmae produced. *Nostoc* colonies visible as dark spots on surface

HABITAT
Areas of disturbed soil such as roadside banks, on moist gravel or sand

OPPOSITE | The *Nostoc* colonies on this *Blasia* are visible as dark-green dots on the thallus surface.

RIGHT | One of *Blasia*'s smooth gemmae balls produced at the end of a long receptacle.

BELOW | *Blasia pusilla* showing both the spiky green gemmae on the thallus surface and also two rows of rounded bumps housing the male antheridia.

NEOHODGSONIA

The Neohodgsoniales is one of several orders of complex thalloid liverworts that are broadly defined as having a differentiated thallus with an upper layer with pores and air chambers and a lower layer of solid tissue. Rhizoids are abundant on the underside of the thallus and may occur in two forms: those that attach the plant to its growing surface and those that can form a capillary conducting system for water retention. The underside of the thallus may have rows of flap-like scales that also have a role in water conduction. Many of these complex thalloid liverworts produce unique, elaborate structures to house the sex organs, which resemble small umbrellas and are formed from the vegetative thallus.

The Neohodgsoniales contains just one species, *Neohodgsonia mirabilis*. It differs wildly from any other thalloid liverworts in having a uniquely branched carpocephalum—the umbrella-like structure that bears the female sex organs. It is also unusual in possessing gemmae cups on its surface, a feature found in a few other liverworts such as the common *Marchantia*. Genetic sequence data has confirmed the uniqueness of this species and its status as the sole representative of this order.

Neohodgsonia lacks the detailed distribution data needed to accurately assess its conservation status,

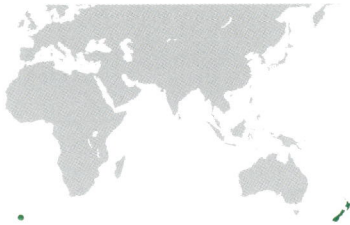

DISTRIBUTION
New Zealand, Tristan da Cunha, Gough Island

ETYMOLOGY
Originally described as *Hodgsonia* after Eliza Amy Hodgson (1888–1983), New Zealand bryologist

NUMBER OF SPECIES
1 known species

APPEARANCE
Thallus dark green with lighter green specks, branches ⅓–⅜ in (8–10 mm) wide, underside with 2 rows of scales. Cells with single large oil bodies. Monoicous. Antheridia and archegonia in separate stalked structures with disc-shaped lobed receptacles. Gemmae cups with frilly margins on thallus surface with discoid gemmae

but it is certainly a rare species known to occur only in New Zealand and the remote Tristan da Cunha islands. Its disjunct distribution is intriguing. In New Zealand it is known to require moist conditions of montane forests since it is not drought tolerant.

OPPOSITE | The round gemmae cups on the thallus surface, similar to those found in *Marchantia*, are just visible in this photo.

RIGHT | The unique branching of the carpocephalum can be seen here.

HABITAT
Usually growing over forest soil in areas of high rainfall

SPHAEROCARPOS

BELOW | *Sphaerocarpos europaeus* belongs to a very distinctive group of thalloid liverworts found on disturbed bare soils.

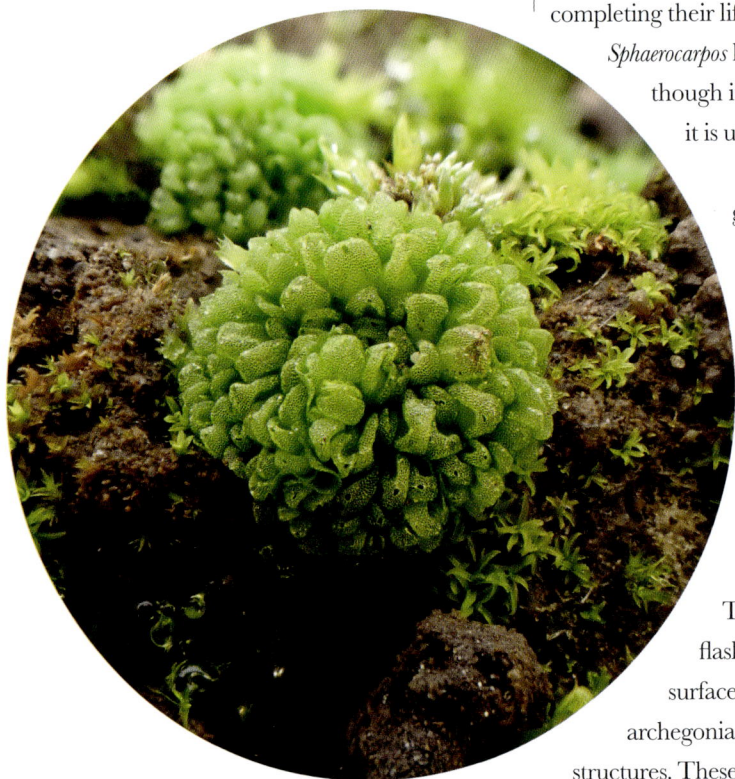

The Sphaerocarpales are sometimes referred to as the "bottle" liverworts in reference to the inflated bottle-shaped protective structures—involucres—that surround the sex organs. They form a very distinctive group of complex thalloid liverworts and are adapted to living in disturbed temporary habitats where they are short lived, completing their life cycle within a year.

Sphaerocarpos has a worldwide distribution, though its range is not continuous and it is usually quite rare.

Sphaerocarpos forms tiny pale-green rosettes with lobed thalli. The dense cluster of balloon-like involucres are extremely distinctive once you get close enough to see them. *Sphaerocarpos* plants are sexually dimorphic, with female plants much larger than the males, sometimes even growing over the male plants. The male antheridia are in the flask-shaped "bottles" on the upper surface of the thallus while the female archegonia are in rounded to cylindrical structures. These plants do not raise their capsules

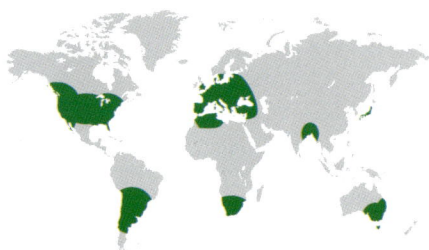

DISTRIBUTION
Pacific North America, South America, Europe, South Africa, southeastern Australia

ETYMOLOGY
Greek *sphaîra* = "sphere" + *karpos* = "fruit," referring to the involucres enclosing the capsules

NUMBER OF SPECIES
10 accepted species

APPEARANCE
Tiny pale-green rosettes with foliose lobes. Dioicous. Male plants only around ³⁄₁₆ in (5 mm) in diameter, antheridia in perigonial involucres on thallus surface. Female plants much larger around ¾ in (2 cm) in diameter, sporophyte enclosed in flask-shaped structure

up on setae like most other liverworts but instead the sporophyte is hidden, completely enclosed within the leafy "bottle" produced from the gametophyte thallus. The capsules do not have lids that open to release the spores; instead, the spores are released when the capsule wall disintegrates. Another anatomical feature of the sporophyte that sets this order apart from other liverworts is the absence of elaters in the capsule, which when present would help to disperse the spore mass. Dispersing spores over long distances is clearly not a priority for *Sphaerocarpos*, and this life strategy may be reflected in the ephemeral habitats that it occupies.

ABOVE RIGHT | A close-up view of *Sphaerocarpos texanus* showing the balloon-like structures that contain the reproductive organs.

RIGHT | *Sphaerocarpos stipitatus*, pictured here in the West Coast National Park, Western Cape, is known only from South Africa and Chile.

HABITAT
On damp soil of disturbed and temporary habitats such as agricultural fields or gardens

ASTERELLA

Marchantiales is a large order containing over 25 genera of complex thalloid liverworts. They have pores on the thallus surface that are surrounded by differentiated cells and an internal thallus structure with air chambers, photosynthesizing cells, and storage tissue. They also have archegoniophores that are tall stalks with a disc-shaped receptacle at the top housing the reduced sporophytes.

Asterella forms dense mats or scattered branches, bright green above and dark wine red to black underneath. Many species of liverworts are highly aromatic, and *Asterella* species can have a characteristic smell of rotten fish.

The archegoniophores in *Asterella* are unique among the complex thalloids. The receptacles are rounded with four or more lobes and each lobe houses a single reduced sporophyte surrounded by a protective sheath or involucre. When the spores are mature this sheath splits into ribbons allowing the spores to disperse. The tattered ribbons are usually conspicuous white or lilac and are a characteristic sight identifying the genus before getting close enough to test for a fishy whiff.

LEFT | The white sheaths surrounding the sporophytes can be seen in this specimen of *Asterella saccata* hanging down below the green receptacles.

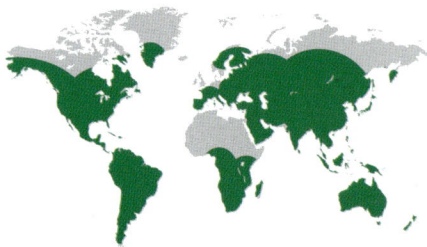

DISTRIBUTION
Worldwide extending from the Arctic, temperate to tropical regions; the Americas, Europe, Africa, Asia, Australasia, and Oceania

ETYMOLOGY
Greek *astēr* = "star," referring to the star-like appearance of the female receptacle

NUMBER OF SPECIES
53 accepted species

APPEARANCE
Thallus green to purple, branches up to ⅜ in (9 mm) wide, underside of thallus with large, often purple scales in 2 rows. Monoicous or dioicous. Antheridia on thallus. Archegoniophores with a conspicuous fringe surrounding each reduced sporophyte

HABITAT
Usually growing on soil, sometimes rock

CONOCEPHALUM

Conocephalum is another example of the complex thalloid liverworts in the Marchantiales with air chambers within the thallus. Looking at the surface of the plants, the borders between the air chambers are clearly outlined with a roughly hexagonal line, giving most species a snakeskin pattern. The openings to the air chambers are via pores that are raised up from the surface like mini volcanoes.

The archegoniophores elongate very rapidly and the stalks holding up the female receptacle can sometimes reach up to 2¾ in (7 cm) high, which is huge for a bryophyte! The female receptacles are conical and a distinctive feature of the genus. Like many of the Marchantiales genera, *Conocephalum* produces aromatic compounds and some species have a strong smell of turpentine when crushed.

These liverworts can form large mats and are commonly encountered in the right kind of damp habitat across the northern hemisphere. Common names are not often applied to bryophytes, yet *Conocephalum* is such a large and easily recognized plant that in English-speaking regions it has acquired two aliases: Great Scented Liverwort and Snakeskin Liverwort.

ABOVE | The clear outline of the air chambers with their central pore gives a distinctive pattern to the surface of this *Conocephalum salebrosum*.

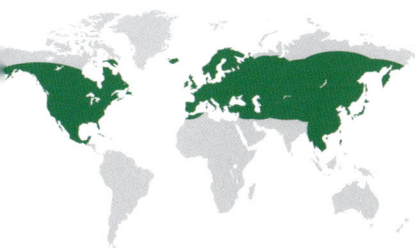

DISTRIBUTION
Predominantly in temperate to boreal regions of the northern hemisphere, spanning North America, Europe, and eastern Asia

ETYMOLOGY
Greek *konos* = "cone" + *kephalos* = "head," alluding to the conical-shaped receptacles

NUMBER OF SPECIES
5 accepted species

APPEARANCE
Large, branched thallus up to 8 in (20 cm) long, surface with raised pores with conspicuous border to air chambers. Dioicous. Archegoniophores produced at ends of thallus branches with long stalks and conical receptacles

HABITAT
Damp areas in woodland, such as near water courses, growing over soil or rocks

LUNULARIA

OPPOSITE | *Lunularia* with its characteristic crescent-shaped gemmae cups housing the tiny green propagules.

BELOW | *Lunularia* is a rather weedy species commonly found around gardens and nurseries where its delightful gemmae cups are always worth a closer look.

The single species representing the Lunulariales, *Lunularia cruciata*, was formerly placed in Marchantiales. However, research into its evolutionary history based largely on DNA sequencing has shown that *Lunularia* is the only living representative of a lineage of plants that diverged from the rest of the complex thalloids early on in their history.

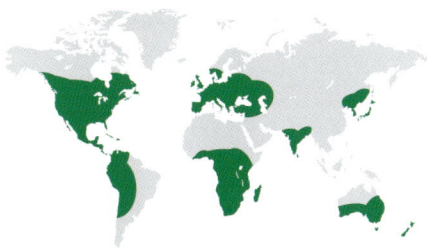

DISTRIBUTION
Worldwide

ETYMOLOGY
Latin *lunula* = "little moon," referring to the crescent-shaped gemma cups

NUMBER OF SPECIES
Only 1 species

APPEARANCE
Thallus bright green, branches up to 2 in (5 cm) long, upper surface with pores, borders to air chambers faintly seen on surface forming reticulate pattern, underside of thallus with 2 rows of scales. Dioicous. Usually sterile. Asexual reproduction by discoid gemmae in lunate gemmae cups on thallus surface

HABITAT
Commonly found in human-made habitats such as gardens or in damp areas near water

Lunularia also differs from the Marchantiales in its anatomical features, particularly the unique crescent-shaped gemmae cups adorning its surface. Lunularia is only occasionally fertile, producing female receptacles with four lobes elevated on a long stalk. Sporophytes are very rarely produced and asexual reproduction is its main means of propagation. Raindrops falling into the gemmae cups splash these tiny green buds away from the parent plant where, if conditions are right, they start to grow.

Lunularia is a rather weedy plant in horticultural nurseries and its numerous gemmae easily spread around into soils. Lunularia has been inadvertently transported all around the world by human activities so its native distribution is rather ambiguous. It is believed to be native to the Mediterranean region, and sexual reproduction has been observed more frequently here than other areas. As Lunularia produces separate male and female plants and disperses widely via its gemmae clones, the chances that only one sex will arrive at a new destination are presumably high. Further research indicates that the rise in annual temperatures may be increasing the frequency of observed sporophytes in some areas, suggesting that climate change is influencing bryophyte reproductive biology.

PELLIA

The round black capsules of *Pellia* are elevated when mature on translucent, ephemeral setae.

Pellia is not the most charismatic of the liverworts and is rather lacking in morphological features. However, it is one of two genera that form the Pelliales order, which represents a distinct lineage of simple thalloid liverworts. There are two genera classified within the Pelliales: *Pellia* and *Noteroclada*, a small genus found in the southern hemisphere that differs from *Pellia* in its more leafy appearance.

The smooth, dark-green, and irregularly branched thalli of *Pellia* plants can form extensive mats with overlapping lobes, sometimes with a purple hint to the thallus. They lack the outlines of air chambers and pores on the surface and the underside of the thallus is without scales, though there are abundant rhizoids.

Pellia can commonly be found in damp, shaded places within its range, and like all bryophytes is dependent on water for fertilization. The antheridia appear as small bumps on the thallus surface, and when mature they absorb water and burst. Sperm cells are then released and in a flurry of frenzied activity must swim to the archegonium,

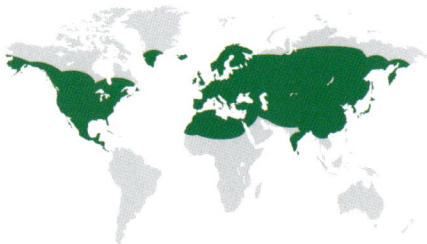

DISTRIBUTION
Widespread across the northern hemisphere in cool temperate regions

ETYMOLOGY
After Leopoldo Pelli-Fabbroni (1783–1822), an Italian lawyer

NUMBER OF SPECIES
9 accepted species

APPEARANCE
Thallus dark green, branches with quite wavy margins, forming mats. Dioicous or monoicous. Sex organs embedded in surface of thallus; antheridia embedded in irregular rows and archegonia in small groups protected by a flap of vegetative tissue. Sporophytes enclosed by protective tube or flap. Spores multicellular and green when mature

fortunately aided by chemical attractants, where fertilization can take place.

Sporophytes can be abundant, with setae sometimes up to 2⅜ (6 cm) tall. When immature the setae are a pale-green color—unusual in liverwort setae, which are mostly translucent. Some species of *Pellia* cannot be identified when sterile because to determine the species it is necessary to know the details of the involucre structure (the flaps and tubes of vegetative tissue protecting the sex organs) and whether the plants are monoicous or dioicous.

ABOVE RIGHT | *Pellia* species are typically found in wet habitats such as in this image where water cascades over mossy rocks.

RIGHT | In fall and early winter, *Pellia endiviifolia* develops numerous branches at the tips, giving it a distinctively frilly appearance.

HABITAT
Damp soil in shaded, wet sites

PETALOPHYLLUM

OPPOSITE | A male plant of *Petalophyllum ralfsii* growing on a dune slack, with the rounded yellow antherida visible.

BELOW | *Petalophyllum* develops separate male and female plants; this female plant of *P. ralfsii* is shown with mature capsules elevated on setae.

*P*etalophyllum is a beautiful, simple thalloid liverwort with erect green sheets of tissue radiating from the middle of the thallus branches to the margin in a regular, ruffled pattern. Across much of its range it is very rare.

The most widespread of the several species of *Petalophyllum* is *P. ralfsii*, which occurs in Europe with a particular stronghold in Ireland. It is adapted to living near the coast in the damp depressions of stable sand dune systems, where it persists from year to year. It may disappear from view entirely when the sand dunes dry out in summer, surviving as underground shoots. Concerns about the declining populations of this charming liverwort caused by factors such as coastal developments damaging its habitat have led to it being among the first bryophytes to gain legal protection at a European level. *P. ralfsii* was thought to also occur in North America until relatively recently, when it was re-evaluated as a separate distinct species, *P. americanum*. This species also favors sandy, seasonally dry soils but it is found in inland habitats rather than coastal sand dunes.

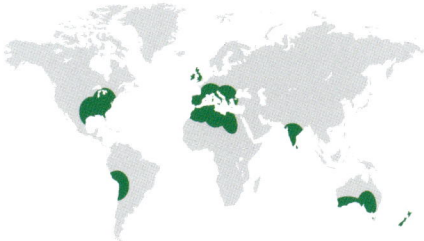

DISTRIBUTION
North America, Bolivia, Europe, Australia, New Zealand

ETYMOLOGY
Greek *petalon* = "petal" or "thin plate" + *phyllon* = "leaf," referring to the leaf-like lobes of the thallus

NUMBER OF SPECIES
5 accepted species

APPEARANCE
Pale-green thallus up to ⅜ in (1 cm) across; upper surface with erect lamellae, underside with 2 rows of scales. Dioicous. Antheridia in rows along central axis of branches, archegonia clustered near branch apex. Sporophytes with various protective structures, which help define the species

HABITAT
Disturbed areas of sandy soil or damp sand dunes

In the southern hemisphere, *P. preissii* of Australia and New Zealand is in even greater peril. Registered as threatened in both countries, there is a real risk that this species may become extinct. A 2014 New Zealand survey found less than 250 individual plants remaining. The exact reason for *Petalophyllum*'s decline here is unclear, though it may be linked to encroachment of its open sandy habitats by invasive grasses.

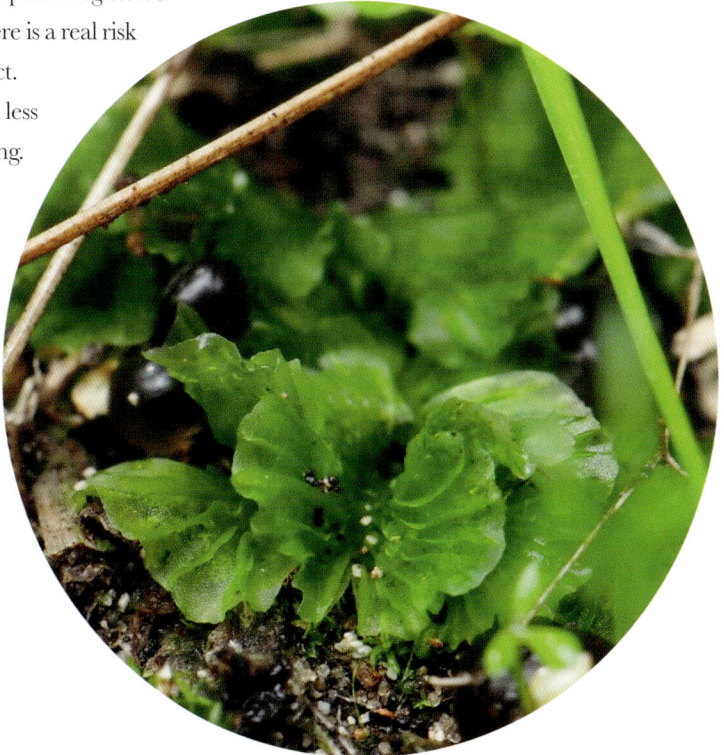

RIGHT | *Petalophyllum preissii*, with its beautiful butterfly-shaped thallus, perilously posed on the brink of extinction.

PALLAVICINIA

OPPOSITE | A female plant of *Pallavicinia lyellii* with frilled scales surrounding a group of archegonia.

BELOW | The elongate thallus of *Pallavicinia lyellii* has a distinctive thickened midrib running down the center.

T his simple thalloid liverwort has linear branches with a prominent thickened midrib that contains small, elongated water-conducting cells in most species. The thallus wings either side of the midrib are noticeably translucent at only one cell thick. The most eye-catching feature of this genus is the distinctive frilly little scales that protect the male antheridia and the long tubes that

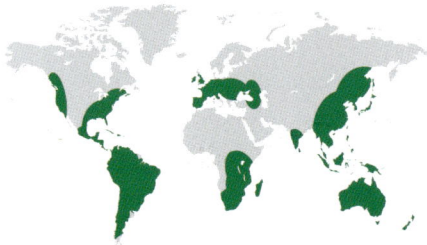

DISTRIBUTION
Widespread

ETYMOLOGY
Probably after Lazzaro Opizio Pallavicini (1719–1785), Italian botanist and archbishop of Genoa

NUMBER OF SPECIES
15 accepted species

APPEARANCE
Simple thallus branches up to 2⅜ in (6 cm) long with well-defined midrib, presence of long stalked slime hairs on thallus margins. Dioicous. Antheridia and archegonia associated with specialist scales. Sporophytes surrounded by long tubular protective structure, capsules elongate

surround the developing sporophyte. Long stalked slime hairs at the branch margins that are only visible under the microscope distinguish *Pallavicinia* from other genera in Pallaviciniaceae.

Pallavicinia is the largest genus in the Pallaviciniales, which is most diverse in the tropics of Southeast Asia. Numerous species were described from this region in the past, though many of these have since been found to represent existing species concepts and have been placed into synonymy. *Pallavicinia* has a global distribution occurring in temperate regions of the northern hemisphere and also in the tropics. In tropical regions it occurs as a forest understory plant but in temperate areas it is mostly found in bogs.

This genus includes another liverwort with a troubled future, since *P. lyellii* is threatened with extinction in Europe. *Pallavicinia* has no specialist structures for vegetative reproduction and as a dioicous plant its populations are often composed of only one sex, limiting its options for reproduction. It is therefore more challenging for *Pallavicinia* to disperse into new habitats. With increased degradation of their bog habitats across Europe due to drainage or habitat destruction, *Pallavicinia* populations have shown a significant decline.

HABITAT
Understory of tropical rainforest, on rotten wood or soil. In temperate regions more likely found in bogs

PLEUROZIA

OPPOSITE TOP |
The complex arrangement of lobules in *Pleurozia gigantea* is particularly evident when viewing the shoots from below.

OPPOSITE BOTTOM |
The tropical *Pleurozia gigantea* growing luxuriantly on an epiphyte-covered tree branch in the montane forests of Réunion Island.

BELOW | The small lobules of *Pleurozia purpurea* can hold onto water, keeping the plant hydrated for longer after rainfall.

Looking more like a cluster of maroon millepedes than a plant, *Pleurozia* is often strikingly colored deep red to purple. The bilobed leaves envelope the stem in two ranks and each leaf is divided into two lobules, one of which is usually smaller and turned into a complex water sac. The sac-like lobe is almost completely enclosed, with a small aperture either side of which there is a flap. This structure is unique to *Pleurozia*. When moist, the flaps close the opening to prevent the water escaping. Microscopic animals have also been observed trapped in this water sac, leading to investigations into potential carnivorous activity.

Given its peculiar morphology, *Pleurozia* has attracted a lot of attention, but it was only in the 1960s that we realized the plant was being described upside down! In the wild the large lobules lie on the upper surface of the leafy shoot and the smaller lobules are hidden underneath. But the large lobules actually constitute the lower side of the stem. *Pleurozia* adapted to keeps its smaller lobules protected underneath the shoot, but when grown in laboratory conditions its orientation is reversed.

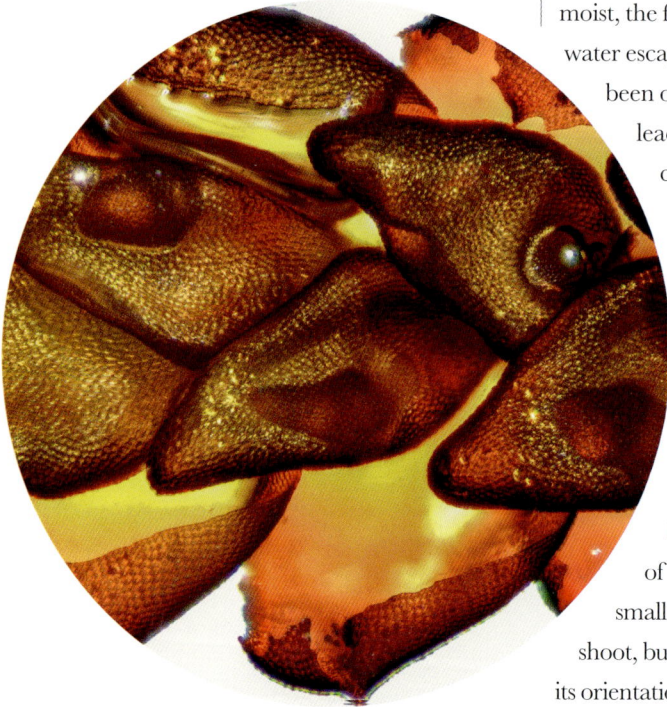

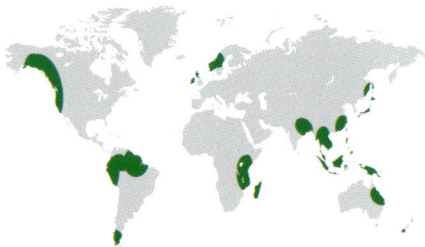

DISTRIBUTION
Widespread but scattered

ETYMOLOGY
Greek *pleuron* = "side" or "rib" + *ózos* = "branch," perhaps referring to the lateral bud-like branches

NUMBER OF SPECIES
12 accepted species

APPEARANCE
Leafy shoots up to 2 in (5 cm) long, green/red/purple. Leaves incubous, bilobed with smaller lobule forming water sac. Underleaves lacking. Dioicous. Sex organs produced on short lateral branches

DNA sequencing delivered another surprise: *Pleurozia* is more closely related to the simple thalloid liverworts than to other leafy liverworts! There are also some clues in the apical cell structure. In *Pleurozia* the apical cells are two-sided, as they are in the simple thalloids, whereas in the true leafy liverworts these are four-sided. This suggests that the morphology of leaves with lobes and lobules evolved independently in *Pleurozia* and the rest of the leafy liverworts.

HABITAT
Epiphytes in tropical montane forests, with one species, *Pleurozia purpurea*, found terrestrially in boreal montane bogs or paramos

METZGERIA

The most commonly encountered member of the Metzgeriales is *Metzgeria*, the thin green ribbons of which can be encountered across the world. The delicate forking branches with a conspicuous midrib line running along the center are immediately recognizable. The plants are typically a bright yellow green when fresh. Interestingly, some species can become bluish when the plants are dried.

For such a simple body plan, *Metzgeria* has diversified into a surprisingly high number of species. Characters to distinguish different species include the position of marginal gemmae; the marginal hairs, single or paired; and whether the plants are monoicous or dioicous. Frustratingly for the taxonomist, these characters are not always consistent. For example, a species may usually have paired hairs at the thallus margin but occasionally single hairs will form.

There is one particularly lovely species, *M. pubescens*, whose narrow lobed thalli are completely coated in tiny hairs, giving the whole plant a plush felt feel. The epiphytic habit of most *Metzgeria* species distinguishes them from other thalloid liverworts that tend to grow over soil or rocks.

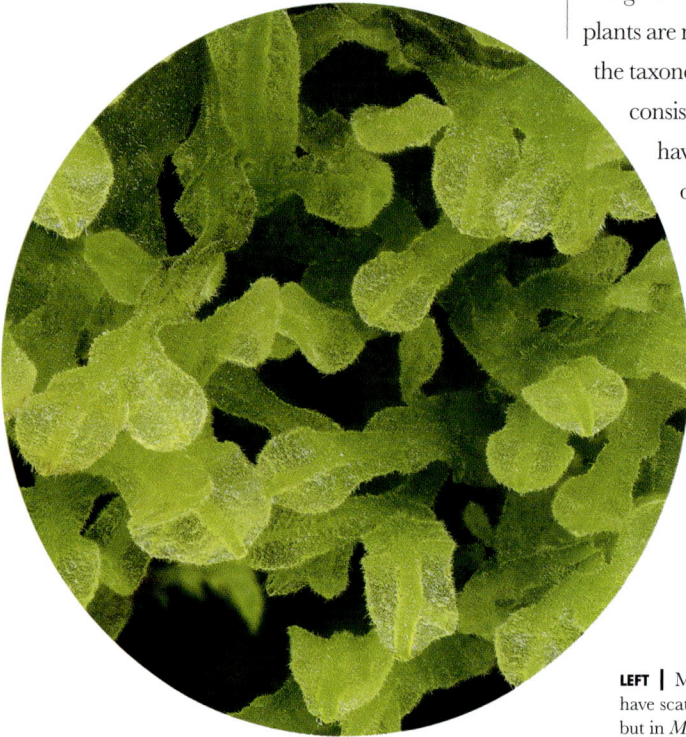

LEFT | Most *Metzgeria* species have scattered hairs on the thallus but in *Metzgeria pubescens* the thallus is delightfully furry.

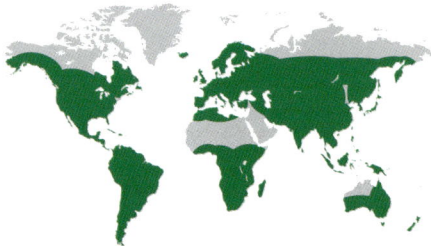

DISTRIBUTION
Worldwide, though most diverse in the tropics

ETYMOLOGY
After Johann Metzger (1771–1844), a copper engraver

NUMBER OF SPECIES
103 accepted species

APPEARANCE
Pale-green, narrow thallus, 1 cell layer thick, dichotomously branched, ³⁄₁₆–6 in (0.5–15 cm) long and ¹⁄₁₆–¹⁄₈ in (2–3 mm) wide, prominent midrib line running along the center, usually with hairs. Monoicous or dioicous. Archegonia and antheridia produced on underside of midrib, sporophyte in a fleshy, often hairy protective covering

HABITAT
Usually epiphytic or occasionally on rocks

ANEURA

T he dark-green creeping thallus branches of *Aneura* are brittle and irregularly branched with no distinct midrib and a characteristic greasy-looking texture. Molecular work analyzing the DNA of these plants is revealing more diversity than has been previously recognized and has also helped redefine the generic boundaries.

A. *mirabilis*, which until recently was classified within its own genus, *Cryptothallus*, is a unique bryophyte that has adapted to a parasitic lifestyle. The plant has no chloroplasts, leaving it a ghostly white throughout. It grows underground, buried up to 8 in (20 cm) deep in damp peat and mats of other bryophytes such as *Sphagnum*. A. *mirabilis* gains its nutrients from a fungus that grows inside the thallus and that is also in a mycorrhizal relationship with a host tree, usually *Pinus* or *Betula*. "Mycorrhizal" refers to the mutually beneficial relationships that fungi can have with plant roots, where the fungi help the roots absorb water and exchange nutrients. These underground invisible networks of living organisms are fascinatingly complex and there is still more to discover about these relationships.

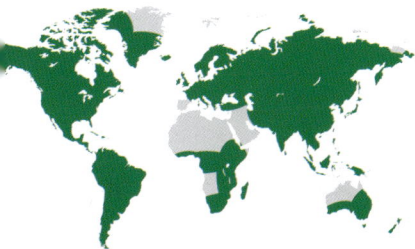

ABOVE | This ghostly white *Aneura mirabilis*, shown here with young sporophytes, was found hidden beneath *Sphagnum* in wet woodland.

DISTRIBUTION
Widespread, highest diversity in wet temperate and subtropical areas

ETYMOLOGY
Greek *an* = "lacking" + *neuron* = "nerve," alluding to the lack of a midrib in the thallus

NUMBER OF SPECIES
54 accepted species

APPEARANCE
Thallus dark green, brittle, greasy, irregularly branched. Up to 2 in (5 cm) long. Lacking midrib. Numerous oil bodies in cells. Dioicous. Antheridia in linear arrangement on thallus. Archegonia at thallus margin. Lacking vegetative reproduction

HABITAT
In moist areas on rocks or soil or among wet vegetation

PTILIDIUM

The Ptilidiales represent one of the earliest diverging lineages within the leafy liverworts. It is a small but well-defined order containing just three families and four genera. These plants are united by several morphological features—for example, they have rhizoids arranged in bundles up the stem originating from the underleaf base; they share a specific style of branching; and they have capsules that are ovoid to elliptical in shape, which is unusual since most liverwort capsules are round.

Ptilidium are handsome plants often tinged with red/orange/green hues and with the lobed leaves finely divided at the margins with a fringe of long cilia. They lack the water sac structures common in the complex lobes of many of the leafy liverworts.

Ptilidium has an interesting global distribution, being widespread across the northern hemisphere but also with isolated ranges such as in southern South America and New Zealand. There has been past debate about whether this distribution pattern is due to chance long-distance dispersal events or whether the isolated southern hemisphere populations represent separate lineages that date back to an early formation of continents in geological history. Current evidence from molecular DNA analysis shows that these southern outliers are closely related to the northern hemisphere plants, suggesting that these populations were sourced from the dispersal of plant fragments.

LEFT | A single leaf of *Ptilidium ciliare* fringed with extravagant long cilia.

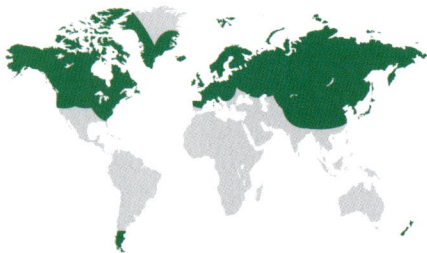

DISTRIBUTION
Across the northern hemisphere, also New Zealand and southern South America

ETYMOLOGY
Greek *ptílon* = "feather," in reference to the feathery appearance of the finely segmented leaves

NUMBER OF SPECIES
4 accepted species

APPEARANCE
Leafy shoots pinnate to bipinnate, leaves incubous, lobed with conspicuously long marginal cilia. Underleaves bilobed. Dioicous. Antheridia and archegonia produced at ends of elongated leafy shoots. Sporophyte rarely produced. Gemmae lacking

HABITAT
Epiphytic on trees, rotting logs, and rocks, or terrestrial growing on soil

LEFT | A rather hairy-looking *Ptilidium californicum* growing on a rotten conifer trunk in British Columbia, Canada.

BELOW RIGHT | A robust leafy liverwort often in shades of yellow and reddish brown, *Ptilidium ciliare* is a widespread species in a range of habitats.

ABOVE | *Ptilidium pulcherrimum* typically grows as an epiphyte on tree trunks or rotting wood.

PORELLA

*P*orella was first described by Dillenius in his *Historia Muscorum* publication of 1741, though it was included within the mosses and only later recognized to be a liverwort. Forming rather robust plants, *Porella* is among the largest of the leafy liverworts. Its branching pattern is distinctively pinnate and it can grow into extensive wefts over tree trunks and rocks. The leaves are divided into a large upper lobe and a smaller lobule that is elongate and held parallel to the stem in a characteristic way. Unlike many other leafy liverworts, this smaller lobe is flat and not formed into a water sac. The reproductive structures in *Porella* are formed on specialist branches. The male

BELOW | *Porella canariensis* in the subtropical laurel forests of Madeira, with a dehisced sporophyte capsule at the base of the shoot.

RIGHT | Most *Porella* species, like this *P. arboris-vitae*, can form large, conspicuous mats over rocks or tree trunks in optimal conditions.

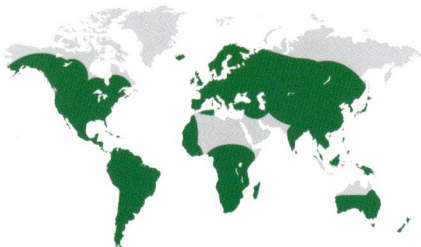

DISTRIBUTION
Widespread

ETYMOLOGY
Latin *porus* = "pore," since Dillenius mistakenly believed the spores were distributed through pores in the side of the capsules

NUMBER OF SPECIES
87 accepted species

APPEARANCE
Robust green-brown plants 2–8 in (5–20 cm) long with pinnate branching. Leaves incubous, divided into a large dorsal lobe and a small ventral lobule, margins entire or toothed. Underleaves lobed. Dioicous, sex organs produced in special branchlets. Sporophytes with large flattened perianths produced on underside of shoots

plants produce stumpy branches off their main shoots reminiscent of tiny catkins and the antheridia are enclosed within the overlapping leaves. On the female plants, sporophytes develop within the protective flaps of what is known as the perianth, which is a particularly large and flattened structure in *Porella*.

Porella is known to produce a range of interesting chemicals such as terpenoids and aromatic compounds, some of which have been shown to have antimicrobial properties. It is not usually recommended to eat the plants one is trying to identify but if one were to have a nibble of *P. arboris-vitae*, for example, it would reveal its identity with its hot and peppery taste.

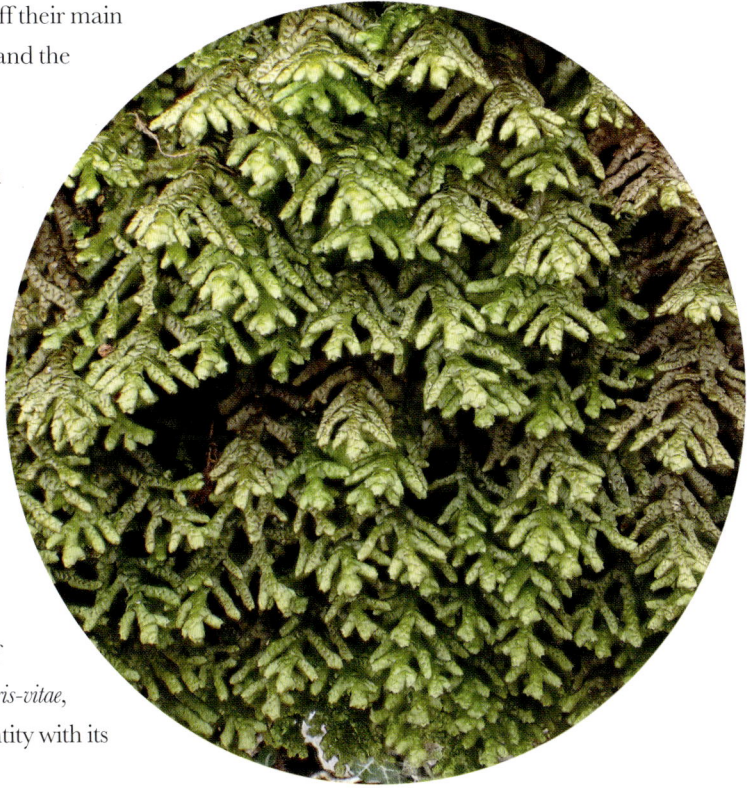

ABOVE RIGHT | A handsome stand of *Porella platyphylla*, demonstrating the typical branching pattern of the genus.

RIGHT | Detail of *Porella cordaeana*, showing the large, broad underleaves and flat, elongated lobules.

HABITAT
On rocky surfaces or epiphytic

RADULA

*R*adula is one of the most diverse genera of leafy liverworts with nearly 250 species. The genus is clearly defined but several attempts have been made to subdivide *Radula* into smaller sections to help describe its exceptional diversity. More recently DNA data has helped resolve some of the more controversial taxonomic issues and provide some useful structure to the large genus.

Different groups of leafy liverworts have consistent patterns in the way their new shoots grow from older branches, and these features provide very important taxonomical characters. *Radula* has a particular type of branching in which a new branch arises in front of the leaf base on the underside of the stem. One of the unique morphological features shown by *Radula* is the origin of the rhizoids, which when present grow directly on the lower leaf lobules rather than the stem. It also lacks underleaves, which

BELOW RIGHT | A single shoot of *Radula complanata*, showing the large, rounded, overlapping lobes.

BELOW | *Radula aquilegia* is a species restricted to highly humid and sheltered habitats, where it typically grows on damp rocks.

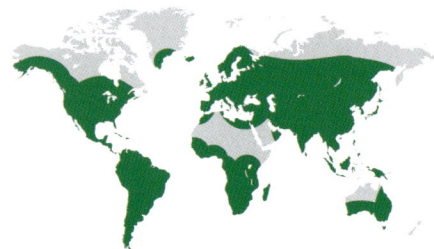

DISTRIBUTION
Widespread

ETYMOLOGY
Latin *radula* = "scraper," referring to the flattened, truncate perianths resembling a scraping tool

NUMBER OF SPECIES
244 accepted species

APPEARANCE
Leafy shoots forming mats, green to brown, ⅜–4 in (1–10 cm) long. Leaves incubous, divided into large dorsal lobe and small ventral lobe. Cells with large, brown oil bodies. Underleaves lacking. Dioicous, rarely monoicous. Perianth narrowly elongate and flattened above with a broad mouth. Gemmae occasional

is unusual for leafy liverworts. The rounded leaves of *Radula* overlap along the stems, giving it a characteristic look.

As with many of the liverworts, *Radula* contains a range of interesting chemical compounds, some with useful pharmacological activity, including cannabinoid-like products. With their ancient origins as early land plants, liverworts have needed to adapt to the harsh conditions of life on land, dealing with challenges such as ultraviolet radiation, desiccation, and an increased threat of herbivory, without making huge investments in robust structural protections. Adaptations toward developing a cocktail of chemicals may have offered some advantage in the face of these challenges.

ABOVE RIGHT | *Radula ankefinensis* is a tropical species growing here in the montane tropical cloud forests of Réunion Island.

RIGHT | Species of *Radula* have cells with characteristic large brown oil bodies, except for *Radula obtusiloba*, which has several large clear oil bodies in each cell.

HABITAT
Usually on tree trunks or rocks, on leaves in humid conditions

FRULLANIA

Until recently *Frullania* was classified within the Porellales order, but compelling evidence from molecular DNA sequencing has shown that this genus is sufficiently distinct to be elevated to its own order.

Frullania is a characteristic epiphyte growing on trees or sometimes rocks in a wide range of habitats around the world, from very dry regions to tropical forests. The plants are often a striking red or purple and the leafy shoots cling tightly to tree bark or form looser wefts. The leaves are divided into two, almost to the base. The smaller lobe lies folded underneath the stem and is characteristically helmet-shaped and arranged in parallel orientation to the stem. This smaller lobe is very variable and the precise shape and size is an important feature for identifying species. A row of underleaves run up the stem, usually forked at the apex and sometimes with teeth at the margin. Sometimes the leaves have special cells called ocelli that stand out since they are slightly larger than the other cells and contain one or more large, dark oil bodies. They might be scattered across the leaf or arranged in a line and are useful taxonomic characters to identify different species.

This extremely large genus has over 600 species and is divided into many subgenera and sections. Most of the diversity is found in tropical regions, with a staggering number of species. The tropical island of New Guinea has the richest *Frullania* flora in the world with an estimated 54 endemic species, which are found nowhere else in the world.

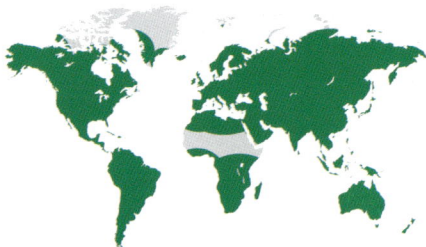

DISTRIBUTION
Globally widespread

ETYMOLOGY
After Leonardo Frullani (1756–1824), Italian politician

NUMBER OF SPECIES
609 accepted species

APPEARANCE
Small to large leafy plants, pinnately branched, ⅜–8 in (1–20 cm) long, red/purple/brown. Leaves incubous, divided with smaller lobule held parallel to stem usually forming a sac. Underleaves bifid. Dioicous or monoicous. The fused leaves that form the perianths have a beaked mouth

ABOVE | *Frullania franciscana* is a species of western North America, shown here on a tree trunk in British Columbia.

ABOVE RIGHT | The smaller leaf lobules of this *Frullania fragilifolia* are distinctly large and helmet-shaped.

RIGHT | *Frullania tamarisci* forming loose reddish-brown wefts over rocks is common across upland areas of Europe.

OPPOSITE | *Frullania* species are often pigmented with reddish-purple hues, as in this brightly colored *F. tamarisci*.

HABITAT
Mostly epiphytic on tree bark and even on living leaves in tropical areas but can also grow on rock surfaces

JUBULA

OPPOSITE ABOVE | *Jubula* tends to grow in characteristic fan-like wefts sticking out from the rocks, as shown by this population.

BELOW | *Jubula hutchinsiae* is found in humid wooded habitats such as this site in the French Pyrenees, usually near streams or waterfalls.

OPPOSITE BELOW | The underside view of a *Jubula hutchinsiae* shoot, showing the spiky, holly-like leaf lobes and large underleaves.

The Jubulales are another recently established order that has been split from the Porellales based largely on molecular DNA evidence. Morphologically these two orders have a lot of similarities, though *Jubula* is green without any of the purple hints characteristic of *Frullania*. Another useful distinction is that *Jubula* often has toothed leaves whereas *Frullania* leaves always have smooth

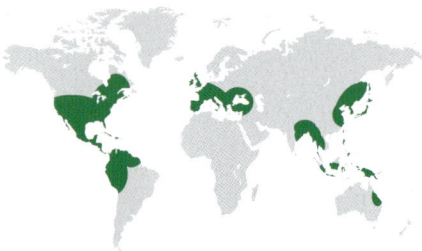

DISTRIBUTION
Widespread in the northern hemisphere and tropics

ETYMOLOGY
Latin *iuba* = "mane" or "crest," referring to features of the spore release

NUMBER OF SPECIES
5 species

APPEARANCE
Green leafy plants, regularly pinnate, around 1½–3⅜ in (4–8 cm) long, leaves incubous, margins sharply toothed or entire, leaf lobule small, forming a sac or flattened. Underleaves bifid with long decurrent bases. Dioicous or monoicous. Perianth with 3 sharp keels and beaked mouth

HABITAT
Soil or rocks, more rarely epiphytic, humid habitats often near flowing water

edges. The smaller leaf lobe usually forms a very small water sac that is held a little way away from the stem. In contrast to *Frullania*, *Jubula* is a small genus with only a handful of recognized species.

The first species of what we now recognize as *Jubula* was described under the name *Jungermannia hutchinsiae* in 1812 by William Hooker, an English botanist who later became the first director of the Royal Botanic Gardens, Kew. Hooker based his description of this new species on plant material collected in Ireland by Ellen Hutchins, who is now celebrated as Ireland's first female botanist. Hutchins lived in a remote part of Ireland called Bantry Bay, where through careful observation and fieldwork she found many interesting and novel plants. She sent plant samples and corresponded with the botanists of the day, especially the English botanist Dawson Turner, whose son-in-law was William Hooker. Tragically Hutchins suffered poor health throughout her life and after just 8 years of

botanizing near her family home died at 29. *Jubula hutchinsiae* is one of several species of plants named after her. It is also one of the most beautiful and a fitting tribute to her memory.

LEJEUNEA

BELOW | A shoot of *Lejeunea cavifolia* with the large, pleated perianth puckered at the end into a short beak.

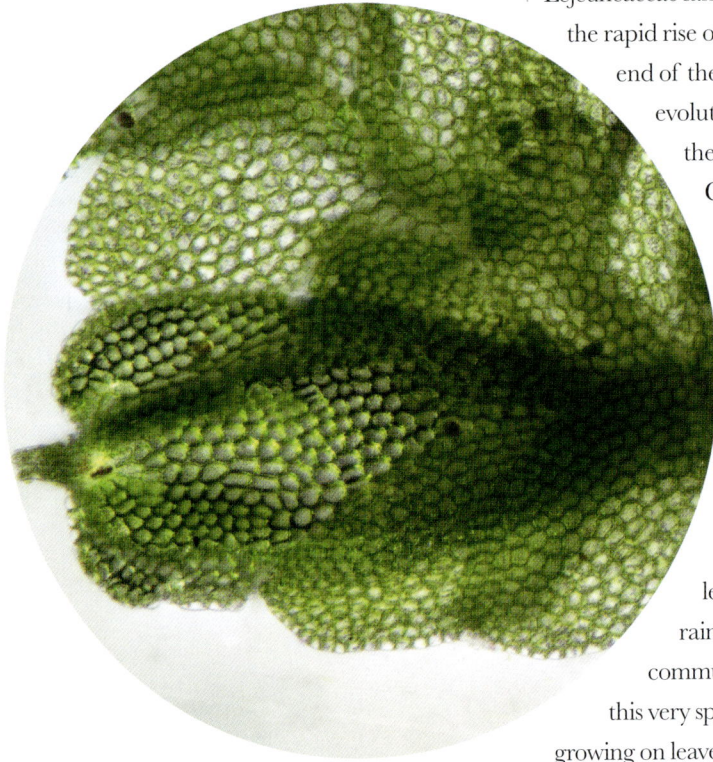

The Lejeuneales are characterized by bilobed leaves with the ventral lobe attached to the dorsal lobe along the ventral margin. This forms a keel at an oblique angle to the stem, which forms an inflated water sac. It is species-rich, with nearly 2,000 species recognized in around 70 genera. A dramatic speciation event among the Lejeuneaceae family can be correlated with the rapid rise of the angiosperms at the end of the Cretaceous period. The evolutionary origins are thought to be in the tropical forests of what is now Central and South America. Within this family, *Lejeunea* is a huge genus of very tiny plants and is still being defined by ongoing taxonomic work.

Plants that specifically grow on the leaves of other plants are called epiphylls. In temperate regions only a handful of organisms can survive growing on leaves, but in humid tropical rainforests, rich and diverse communities of tiny liverworts thrive in this very specific habitat. Almost all plants growing on leaves in tropical forests belong to the

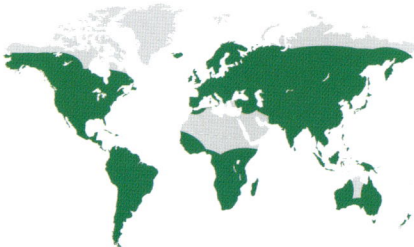

DISTRIBUTION
Globally widespread

ETYMOLOGY
After Alexandre Louis Simon Lejeune (1779–1858), a Belgian botanist

NUMBER OF SPECIES
410 accepted species

APPEARANCE
Plants tiny to robust, up to 4 in (10 cm) long and ⅛ in (3 mm) wide, pale green or whitish, never reddish. Leaf

margin rarely toothed. Underleaves bifid. Dioicous or monoicous. Sex organs on short specialist branches, perianths with 0–5 keels, variously ornamented. Gemmae lacking

HABITAT
Mostly epiphytic on tree trunks, branches, and leaves, less commonly on rocks

Lejeuneaceae. Epiphylls are particularly vulnerable
to forest disturbance since they are completely
dependent on the microclimate created by the thick,
closed canopy of forest trees. These communities
can be similarly impacted by
the introduction of non-native tree species, which
can become invasive and change the forest
composition, affecting the level of moisture in the
air. Conservation measures for these diverse
communities of leaf-dwelling liverworts rely on
the safeguarding of their forest habitats.

COLURA

The tiny pale green cushions of *Colura* are easily recognized under the magnifying glass. The miniscule leaves form an inflated lobule ending in a long beak-like tube. This lobule has an opening and closing mechanism formed by a valve and a hinge, and the morphology of the valve is the basis for dividing *Colura* into several taxonomic sections.

Two liverworts, *Colura* and *Pleurozia* (Pleuroziales), have been investigated for a rather surprising adaptation in their lobules. These genera are not closely related yet both share a trap-like mechanism in the lid of modified leaves, forming water sacs. Microscopic animals such as ciliates that feed on bacteria growing on the leaves of liverworts have been observed trapped inside these water sacs—once they enter, the lid closes and cannot be opened from the inside. There, the inevitable awaits . . . Once the animal dies, it is soon digested by bacteria, and nutrients are potentially available to the plant. Are these scenes evidence of a carnivorous lifestyle? That small animals become trapped in these water sacs has been observed both in the wild and in laboratory experiments. Whether the liverwort actually makes use of the nutrients released by the decomposition of its "prey" has not been substantiated, and so whether these plants can be considered carnivorous is not yet known. We can be sure, though, that these water sacs do perform a very elegant solution to a challenge of holding water close so the plant cells can remain metabolically active in between rain showers.

LEFT | A single shoot of *Colura calyptrifolia* showing the inflated lobes with a long beak, resembling tiny teapots.

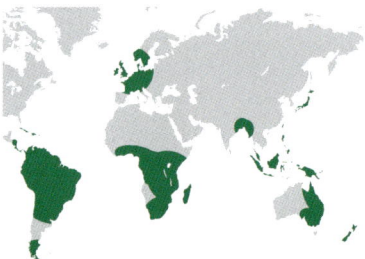

DISTRIBUTION
Widespread in the tropics, some extensions into northern and southern hemispheres

ETYMOLOGY
Greek *kólos* = "docked" or "hornless" + *ourā* = "animal tail," referring to the perianth in *Colura calyptrifolia* with its 5 keels resembling stunted horns

NUMBER OF SPECIES
86 species

APPEARANCE
Plants tiny, up to 1/16 in (2 mm) wide, pale yellow green. Leaves longer than wide, upper half forming water sac with valve-hinge mechanism. Underleaves deeply divided. Perianths keeled, sometimes with horned ends

BELOW | There are a huge number of described *Colura* species, often differing in tiny details of the lobule structure. This *Colura tenuicornis* is forming a dense colony on a river boulder on Table Mountain, Western Cape, South Africa.

ABOVE | A single leaf of *Colura calyptrifolia* with a small lobe surmounted by a larger lobule and ending in a long beak. Small creatures are often trapped in the inflated lobe, like the tardigrade inside this one, visible as the brown shape to the lower right of the picture

HABITAT
Epiphytic, mostly epiphylls in tropical forests; rarely on peat bogs

SCHISTOCHILA

These handsome plants in shades of green or red are some of the largest and most beautiful liverworts. They can be a characteristic element of humid montane forests within their distribution range, which is restricted to the southern hemisphere. Distinctive features that set *Schistochila* apart from any other of the larger leafy liverworts include the rather flattened appearance of the shoots and the distinctly elongate leaves, which are at least twice as long as wide. These are folded lengthways to form two lobes connected along a winged keel. The leaf margins can be toothed or ciliate. The developing sporophyte is surrounded by a swollen, fleshy sleeve called a perigynium, whereas in other genera different structures protect the sporophyte.

Schistochila produces numerous rhizoids that can be colorless or a striking red-magenta color. They can also be highly branched, with numerous cell walls dividing the rhizoids up into segments—an unusual trait in the liverworts. Examination of fresh material recently showed these rhizoids to be filled with fungal hyphae, which was an exciting and unexpected discovery. Some liverwort groups can form associations with endophytic fungi, as seen to extreme effect in *Aneura mirabilis*. A particularly interesting find in *Schistochila* is that it is the fungi that are inducing the liverwort rhizoids to branch and form cross-walls. These findings are increasing our understanding of terrestrial plant–fungal interactions, which is a fascinating area of research given the importance of mycorrhizal associations for plant health.

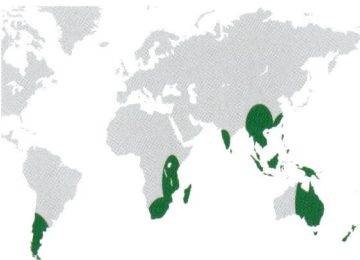

DISTRIBUTION
Mainly southern hemisphere

ETYMOLOGY
Greek *schistós* = "split" or "divided" + *cheílos* = "beak" or "rim," alluding to the fringed rim of the perianth

NUMBER OF SPECIES
90 species

APPEARANCE
Large, leafy, flattened shoots up to around 2 ⅜ in (6 cm) long and ⅜ in (1 cm) wide. Leaves incubous or succubous, bilobed, margins usually toothed to ciliate. Underleaves present or absent. Sex organs on long branches. Sporophyte surrounded by fleshy perigynium. Gemmae lacking

OPPOSITE | *Schistochila balfouriana* showing the large forked underleaves and long cilia.

ABOVE | *Schistochila* are large, attractive plants with a distinctly flattened appearance. They are common in the wet forests of New Zealand where this species, *S. appendiculata*, was photographed.

RIGHT | *Schistochila aligera*, photographed here in the cloud forests of Rarotonga in the Cook Islands.

HABITAT
On tree trunks and rotting logs in humid montane forests

MYLIA

The Myliales are a recently established order described in 2023. It includes just the one genus, *Mylia*, with four species that were previously classified under the broader concept of the Jungermanniales. The two species *M. taylorii* and *M. anomala* are widespread across North America and Europe, while *M. verrucosa* and the recently described *M. vietnamica* occur in East Asia.

The plants can form very beautiful reddish-purple cushions or occur as scattered stems, for example through *Sphagnum*. The nearly round, unlobed leaves are often pressed together at the tip of the shoot and the relatively large leaf cells can be seen with a magnifying glass. The size of these cells can help quickly distinguish *Mylia* from any round-leaved leafy liverwort. The rhizoids are produced in tufts associated with the leaf bases and underleaves. *Mylia* often produce gemmae around the edge of their leaves and in some species these are produced on specialist shoots that have more elongated leaves.

M. taylorii is commonly associated with *Sphagnum* peat bogs across North America and is part of a wonderful community of bryophytes and other plants that live in these fascinating habitats. As well as looking beautiful, fresh plants of *M. taylorii* are said to have an aroma of cedarwood, though these liverwort scents can be hard to detect.

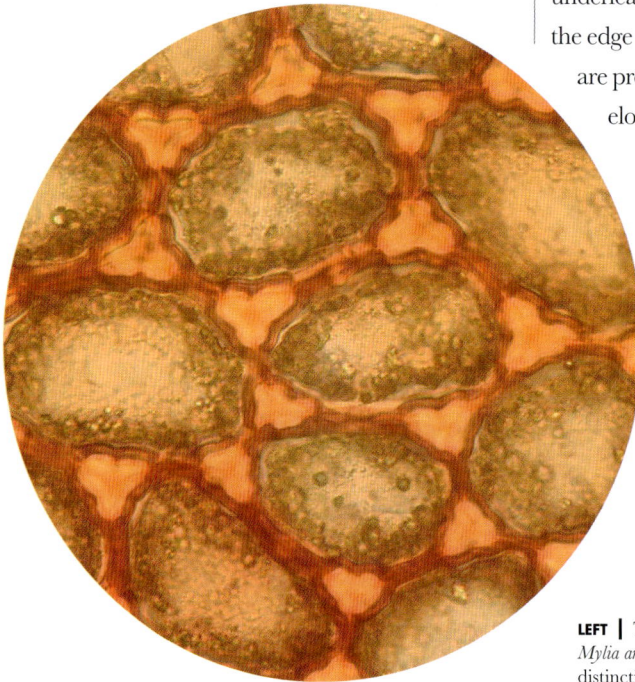

LEFT | The large cells of this *Mylia anomala* leaf also show distinctive cell wall thickenings.

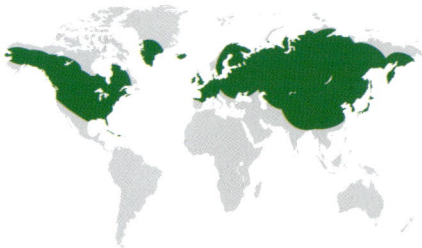

DISTRIBUTION
Restricted to northern hemisphere—North America, Europe, Asia

ETYMOLOGY
After Willem Mylius (1674–1748), a Dutch physician

NUMBER OF SPECIES
Only 4 accepted species

APPEARANCE
Leafy shoots, simple or sparsely branched, forming loose mats, green to red, up to 4 in (10 cm) long and 3/16 in (0.5 cm) wide, leaves succubous, almost circular, leaf cells very large. Underleaves tiny. Dioicous. Sporophyte enclosed in flattened perianth. Gemmae produced on leaf margins

HABITAT
Growing humus, peat, or rotting wood, or rocks

RIGHT | *Mylia taylorii* can form large swelling cushions and the plants are often colored a striking purplish red.

BELOW LEFT | When *Mylia taylorii* grows in deep shade, the plants lack the more typical purplish-red coloration often associated with this species and are instead quite green.

BELOW RIGHT | A single shoot of *Mylia anomala*, with rounded lateral leaves and the tiny underleaves barely visible.

SCAPANIA

The liverworts belonging to the Lophoziales were until recently included in the Jungermanniales. They have, however, been found to represent a sufficiently distinct lineage to warrant recognition at the ordinal level and include over 600 species.

Scapania can usually be identified by its complicated bilobed leaves in which the front lobe is the smaller and overlaps the larger lobe behind. Only occasionally are the lobes a similar size. Rhizoids are scattered along the stem, not in clumps. The perianth structures are distinctively smooth and flattened. The frequent gemmae produced on the leaves can also be a useful character to identify species, since the gemmae are produced in a range of colors, from green to red to brown. It is a large genus with over 100 accepted species. Most of its diversity is centered in the northern hemisphere, though a few species have extended into the tropics.

Some *Scapania* species in temperate regions are commonly associated with aquatic environments, for example growing on rocks in small streams. Several studies on aquatic bryophytes have demonstrated their ability to accumulate heavy-metal pollutants

LEFT | *Scapania americana* is a species endemic to western North America, photographed here in the temperate conifer forests of British Columbia.

DISTRIBUTION
Widespread

ETYMOLOGY
Of uncertain origin; *scapa*, probably related to a Greek word for "dig," possible later meaning of "spade," referring to the spade-like flattened perianth

NUMBER OF SPECIES
107 accepted species

APPEARANCE
Small to rather large plants; green, red, or brown, 3/16–8 in (0.5–20 cm) long, sparsely branched. Leaves bilobed into smaller and large overlapping lobes (occasionally the 2 lobes subequal). Underleaves lacking. Dioicous. Perianths flattened, smooth. Gemmae frequent

from the water into their cell walls. This talent has been observed in *S. undulata*, which seems to have some tolerance to heavy-metal pollution and could provide a useful tool for monitoring levels of pollutants in streams.

HABITAT
Soil, rocks, small streams, decaying wood, less often epiphytic on trees

SYZYGIELLA

The main distinctive feature for the *Syzygiella* has traditionally been the two leaves that join together at the leaf base, like holding hands around a tree trunk. However, the concept of the genus has been expanded slightly since first described and now includes a wider diversity of plants with

opposite and also now alternately arranged leaves up the stem.

Molecular DNA analysis across the liverworts usually serves to split up large genera into smaller, more discrete units with shared evolutionary history. *Syzygiella* is unusual in this respect since molecular evidence here has supported a broader concept of this genus, with *Cryptochila* and *Jamesoniella* recently included within *Syzygiella*. This larger, more inclusive *Syzygiella* can still be recognized by its usual red-purple coloring and inflated plicate perianths.

Syzygiella has its highest diversity in the southern hemisphere and mountainous tropics, with just a few species found also in the temperate northern hemisphere.

LEFT | *Syzygiella colorata* is a widespread species in the southern hemisphere, showing the reddish-purple coloration typical of this genus.

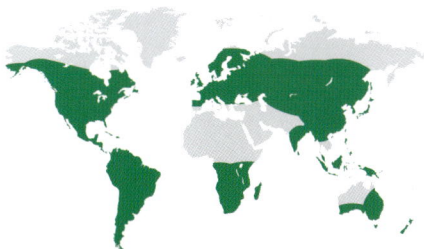

DISTRIBUTION
Widespread

ETYMOLOGY
Greek *súzugos* = "paired" or "joined together," referring to the opposite stem leaves joined at the leaf base

NUMBER OF SPECIES
37 accepted species

APPEARANCE
Plants small to large, creeping or erect with apex of shoot curved over,

⅜–4 in (1–10 cm) long, usually with some red-purple coloration. Leaves succubous, opposite or alternate. Leaves rounded or shallowly bidentate. Underleaves tiny or lacking. Dioicous. Perianths inflated and plicate. Gemmae lacking

HABITAT
Usually epiphytic, or on rotten wood or rocks

GONGYLANTHUS

The Jungermanniales is the fourth-largest liverwort order with over 550 species and 59 genera. It forms a monophyletic group with the Lophoziales, Lepidoziales, Myliales, and Perssoniellales, meaning they share a common ancestor, and until recently these orders were all classified within the Jungermanniales. Interestingly, some leafy liverwort orders such as the Lejeuneales showed a rapid increase in species numbers around the time that angiosperms were forming vast tropical forests since these liverworts were adapted to living on other plants and radiated to exploit new habitat potential. In contrast, the Jungermanniales do not show this same pattern, since they include more terricolous species that inhabit soil and rock surfaces.

Gongylanthus is a genus generally restricted to warmer regions. It is a very pretty plant with rounded leaves arranged in pairs. Its green, leafy, creeping shoots can be recognized by the leaves arranged opposite each other in pairs up the stem with their leaf bases united. It is more common for leafy liverworts to have leaves alternately arranged up the stem, so the opposite arrangement immediately stands out.

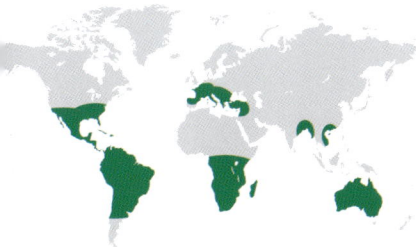

ABOVE | *Gongylanthus ericetorum* creeping over rocks in the subtropical laurel forests of Madeira.

DISTRIBUTION
North and South America, Southern Europe, Africa, eastern Asia, Australia

ETYMOLOGY
Greek *gongylis* = "turnip" or "root vegetable" + *Anthos* = "flower," referring to the marsupium (swollen sac) growing into the ground and protecting the sporophyte

NUMBER OF SPECIES
10 species

APPEARANCE
Leafy shoots usually unbranched, up to 4 in (10 cm) long, green sometimes with purple hints, leaves succubous, opposite, leaf apex rounded, edges smooth. Underleaves lacking or tiny. Dioicous. Sporophytes in fleshy marsupium instead of a perianth

HABITAT
On bare soil or rock

SOLENOSTOMA

In the early nineteenth century most leafy liverworts were classified in the genus *Jungermannia*. At this time the guiding principles in bryophyte classification were weighted heavily toward features of the sporophyte, and leafy liverworts, with their very uniform sporophytes, were largely lumped into one genus. In time,

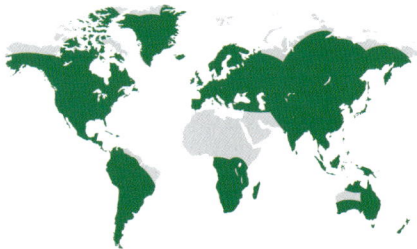

BELOW | *Solenostoma vulcanicola* forming vivid green cushions at Chatsubomi Moss Park in the town of Nakanojō, Japan.

DISTRIBUTION
Widespread

ETYMOLOGY
Greek *sōlēn* = "pipe" + Greek *stoma* = "mouth," referring to the shape of the perianth

NUMBER OF SPECIES
143 accepted species

APPEARANCE
Green, brown, or dark purplish leafy shoots, up to 1½–2 in (4–5 cm) long, rarely branched, leaves succubous, undivided, apex rounded and margins entire. Lacking underleaves. Dioicous. Perianths plicate, suddenly constricting at mouth and forming a tubular beak. Gemmae lacking

HABITAT
Various substrates from rock to soil or tree trunks, occasionally as submerged aquatics

Marchantiophyta: The Liverworts

bryologists began to put more emphasis on the gametophyte diversity, and numerous genera began to be separated out from *Jungermannia* based on distinct leaf forms and cell characteristics. In recent times molecular DNA evidence has resulted in even more changes to the Jungermanniales, and the boundaries between genera in this order are still being defined.

Solenostoma are usually dark-green, leafy plants with undivided rounded leaves and perianths that form a distinctive tubular beak at their tip. Some species of *Solenostoma* associated with aquatic habitats show a tolerance for heavy metals. *S. vulcanicola* flourishes in the sulfuric hot springs and acidic rivers sourced from volcanic origins in Japan. In these extraordinary habitats where few if any other plants can grow, it forms extensive bright-green cushions and is known locally as "Chatsubomi moss." This plant commonly lives in waters with

a pH of 3–4.5 but it has even been observed at Kusatsu hot spring with a pH of 1.9. For context, lime juice usually has a pH of 2–3. Chatsubomi moss is a lesser-known marvel of the region but it does not go unnoticed, with specialist tours available to visit its most impressive sites.

BAZZANIA

Perhaps the most animal-like of all the liverworts, *Bazzania* creeps over rocks and branches like green millipedes on the march. It is classified within the huge order of the Lepidoziales, which are mainly epiphytic habitat specialists with characteristically three- or four-pronged leaf lobes.

BELOW | *Bazzania trilobata* can form distinctive huge hummocks when growing well in humid habitats—here in a mossy woodland in the English Lake District.

Bazzania tends to branch in an irregularly forked Y-shaped pattern. The shoots have a rounded back with the leaves closely overlapping in the incubous manner and with the leaf tips turned downward. The leaves are divided into two to three points at the apex, there are well-developed underleaves that are similarly pronged, and often there are long, thin, flagelliform branches that hang down from the main shoots. *Bazzania* rarely reproduces sexually and the lack of spores may suggest why particular species within this genus are less widely dispersed across continents than is the case in other large tropical liverwort genera.

Although most plants can usually be assigned to *Bazzania* quite easily, identifying species can be notoriously tricky. This is especially true for regions where the genus is particularly diverse, for example in tropical montane forests. One of the challenges is that particular species can grow in slightly different ways depending on their environment, and this can stretch the parameters of specific measurements and shapes that have been used to describe a species. For example, plants growing in moist, shaded habitats may be smaller plants with less well-developed leaves, perhaps with fewer pronged ends, than the same species growing in sunny, exposed spots.

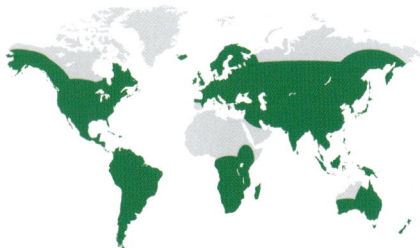

DISTRIBUTION
Widespread

ETYMOLOGY
After Matteo Bazzani (1674–1749), an Italian anatomy professor

NUMBER OF SPECIES
Around 256 accepted species

APPEARANCE
Leafy shoots around 1¼–4 in (3–10 cm) long, yellow green to dark green, dichotomous branching often with long flagella from underside. Leaves incubous, apex divided into 2–3 (sometimes 4) teeth. Underleaves also with apex divided. Dioicous. Sporophytes uncommon and reproduces vegetatively by leaf fragments

HABITAT
Occupies a range of habitats, from tree trunks to logs or rocks in damp sites

ABOVE | The pale green shoots of *Bazzania pearsonii* growing in Scotland with *Pleurozia purpurea*.

ABOVE | The leaves of the *Bazzania denudata* easily fall off, as can be seen with some exposed stems of this plant, photographed here in British Columbia.

RIGHT | Moist shoots of *Bazzania tricrenata* showing the characteristic three- or four-pronged underleaves.

HERBERTUS

The *Herbertus* genus is one of the few liverworts that could be mistaken for a moss at first glance. It has uniquely forked leaves giving the shoots the look more typical of the mosses. The magnifying glass soon reveals deeply divided leaves, though, which the mosses are incapable of developing. The plants usually form upright shoots that are occasionally creeping, and they can get very large, forming cushions up to 12 in (30 cm)

deep in some cases! In tropical montane forests they can form part of the extensive leafy liverwort communities weighing down tree branches with their thick blankets.

As with many of the liverworts, fitting a plant to a genus is usually possible with a bit of practice, but with some of these very large genera, identification to species level can be quite a challenge. *Herbertus* is rarely fertile, so all characters for distinguishing species are mostly based on the leaf shape and the angle of the leaf on the stem. The leaves have a vitta, which is a band of enlarged cells running along the middle of the leaf lobe, and the size and length of this band is also used in species identification. DNA analysis has provided a lot of support in working out species boundaries where it has been applied, but the vast majority of tropical species have not yet been DNA sequenced and these studies also require detailed morphological research to generate a comprehensive taxonomy.

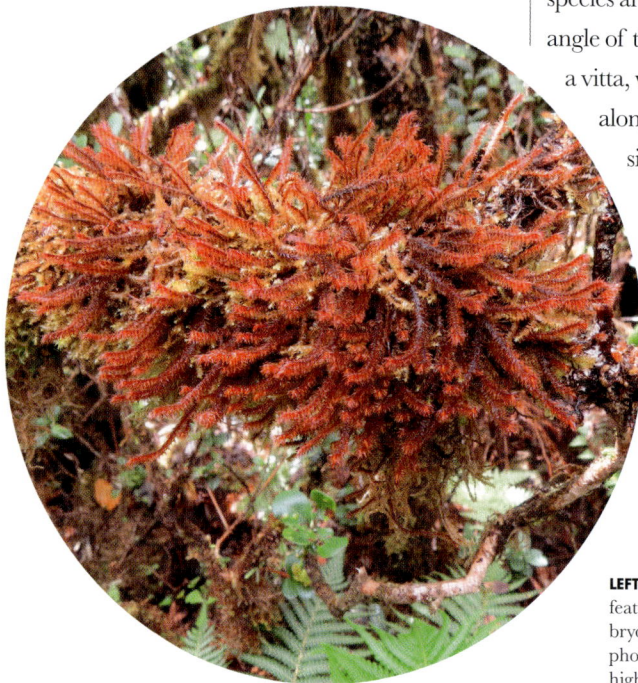

LEFT | *Herbertus* species are a common feature of the tropical montane bryophyte flora, with this plant photographed in the cloud forest highlands of the Hawaiian Islands.

DISTRIBUTION
Widespread, most diverse in tropical/subtropical areas

ETYMOLOGY
After Thomas Herbert (1766–1828), a British nobleman and patron of the sciences

NUMBER OF SPECIES
40 accepted species

APPEARANCE
Plants usually robust, 4–12 in (10–30 cm) high, varied in color from green to almost black to wine red or rusty brown. Leaves usually incubous, deeply forked into 2 lobes with long, acute to fine points, vitta present. Dioicous. Sporophytes rarely produced

HABITAT
Growing as epiphytes, on rocks or forming turfs on the ground in humid sites

LEFT | A leaf of *Herbertus stramineus* showing the well-defined vitta extending well into the leaf lobe.

BELOW RIGHT | *Herbertus* cf. *stramineus* on a wet rock face in the temperate coniferous forests of British Columbia.

BELOW LEFT | *Herbertus borealis* is a Scottish endemic, known only from the Northwest Highlands, a region with a cool and very wet climate.

PLAGIOCHILA

The *Plagiochila* genus was added in 2023 to a redefined Lepidoziales order, reflecting a shared ancestry with this group and a similar time period for when *Plagiochila* and the rest of the Lepidoziales appear to have diverged from the other liverwort orders.

Plagiochila plants are usually robust with upright shoots and can form strikingly large hummocks in ideal conditions. The stems have a slightly thicker surrounding layer than most other liverworts, so they are structurally a little more robust. The most characteristic feature of *Plagiochila* is its succubous leaves (arranged on the stem so that the top edge of a leaf is positioned underneath the leaf above it), which are undivided, slightly swept backward, and with a long, decurrent leaf base running down the stem. The margins are often toothed like tiny holly leaves. The perianths surrounding the developing sporophyte are distinctly flattened and often fringed at the mouth. This basic body plan has produced a bewildering number of species, and *Plagiochila* is the most species-diverse of all the liverwort genera.

There are numerous examples of bryophytes being used in traditional medicine around the world. Liverworts contain a range of biological compounds that could have useful pharmaceutical applications, though the majority have not yet been investigated. *P. beddomei* is one such example, having traditionally been used for healing wounds in the Western Ghats, India. Scientific analysis of extracts from this liverwort has indeed demonstrated a positive impact on the rate at which wounds heal by promoting the growth of new blood vessels.

LEFT | An impressive mound of *Plagiochila heterophylla* in a humid, mossy woodland of North Wales.

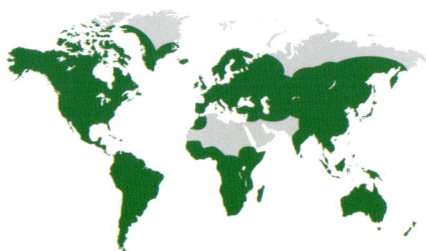

DISTRIBUTION
Worldwide

ETYMOLOGY
Greek *plágios* = "slanting" + *cheílos* = "beak" or "rim," referring to the flattened perianths with wide truncate mouth

NUMBER OF SPECIES
633 accepted species

APPEARANCE
Plants small to large, up to 8 in (20 cm) long, green or brown but never red. Leaves unlobed and succubous, often with strongly decurrent leaf base and recurved dorsal margin, often toothed. Underleaves absent or tiny. Dioicous. Antheridia on long specialist branches. Perianth flattened with a broad mouth

ABOVE | A beautiful shoot of *Plagiochila appalachiana*, which is endemic to the southern Appalachian region of the USA.

RIGHT | *Plagiochila semidecurrens* on a wet rock face in the temperate coniferous forests of British Columbia.

HABITAT
Epiphytic, over soil or on rocks

BRYOPHYTA: THE MOSSES

Most mosses have leaves spiraled around their stems and robust setae elevating their capsules. The peristome structure is a unique and sophisticated feature of the mosses, though this innovation has been lost or reduced in many lineages. The architecture of the peristome has traditionally played a fundamental role in defining major lineages of mosses. For what might seem a rather basic body plan, mosses show a staggering diversity of forms. There are an estimated 12,500 species of mosses classified across 44 orders and 969 genera.

The Takakiales have segmented leaves and spirally dehisced capsules. Several features set the Sphagnales apart from other mosses, such as the dimorphic leaf cells and distinct sporophytes. The Andreaeales and Andreaeobryales have unusual, valved capsules, while the Polytrichales and Tetraphidales are defined by their "nematodontous" peristomes, meaning the teeth are made of whole dead cells. Most of the remaining moss orders fit into the large diverse Bryopsida class, broadly defined by an "arthrodontous" peristome, meaning the teeth are made of parts of dead cells. Within the diverse Bryopsida, over half of the species diversity lies within the orders of creeping pleurocarpous mosses, which speciated rapidly alongside the emergence of the tropical angiosperm forests.

PHYLOGENETIC TREE OF MOSS ORDERS AND FEATURED GENERA

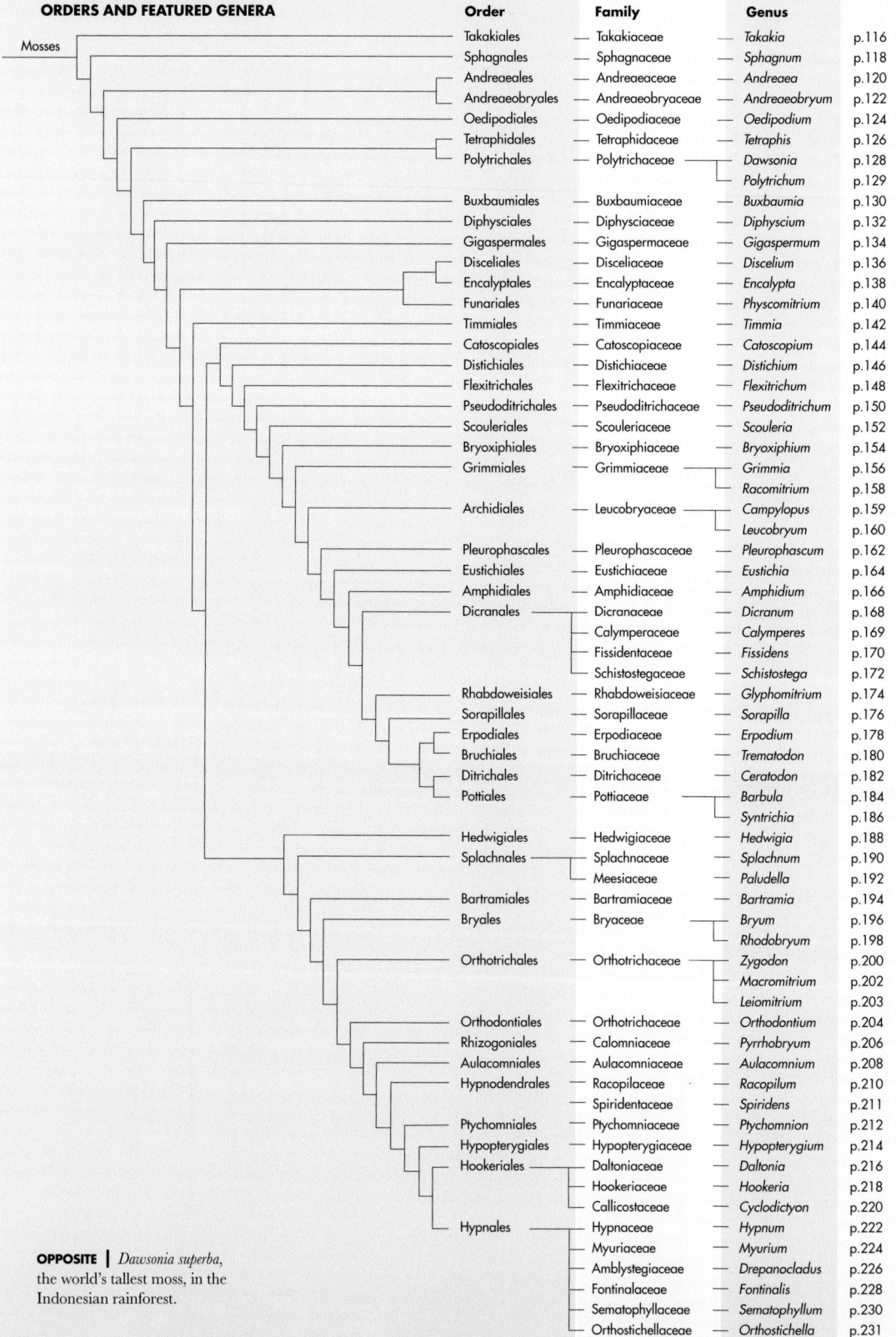

Mosses

Order	Family	Genus	
Takakiales	Takakiaceae	Takakia	p.116
Sphagnales	Sphagnaceae	Sphagnum	p.118
Andreaeales	Andreaeaceae	Andreaea	p.120
Andreaeobryales	Andreaeobryaceae	Andreaeobryum	p.122
Oedipodiales	Oedipodiaceae	Oedipodium	p.124
Tetraphidales	Tetraphidaceae	Tetraphis	p.126
Polytrichales	Polytrichaceae	Dawsonia	p.128
		Polytrichum	p.129
Buxbaumiales	Buxbaumiaceae	Buxbaumia	p.130
Diphysciales	Diphysciaceae	Diphyscium	p.132
Gigaspermales	Gigaspermaceae	Gigaspermum	p.134
Disceliales	Disceliaceae	Discelium	p.136
Encalyptales	Encalyptaceae	Encalypta	p.138
Funariales	Funariaceae	Physcomitrium	p.140
Timmiales	Timmiaceae	Timmia	p.142
Catoscopiales	Catoscopiaceae	Catoscopium	p.144
Distichiales	Distichiaceae	Distichium	p.146
Flexitrichales	Flexitrichaceae	Flexitrichum	p.148
Pseudoditrichales	Pseudoditrichaceae	Pseudoditrichum	p.150
Scouleriales	Scouleriaceae	Scouleria	p.152
Bryoxiphiales	Bryoxiphiaceae	Bryoxiphium	p.154
Grimmiales	Grimmiaceae	Grimmia	p.156
		Racomitrium	p.158
Archidiales	Leucobryaceae	Campylopus	p.159
		Leucobryum	p.160
Pleurophascales	Pleurophascaceae	Pleurophascum	p.162
Eustichiales	Eustichiaceae	Eustichia	p.164
Amphidiales	Amphidiaceae	Amphidium	p.166
Dicranales	Dicranaceae	Dicranum	p.168
	Calymperaceae	Calymperes	p.169
	Fissidentaceae	Fissidens	p.170
	Schistostegaceae	Schistostega	p.172
Rhabdoweisiales	Rhabdoweisiaceae	Glyphomitrium	p.174
Sorapillales	Sorapillaceae	Sorapilla	p.176
Erpodiales	Erpodiaceae	Erpodium	p.178
Bruchiales	Bruchiaceae	Trematodon	p.180
Ditrichales	Ditrichaceae	Ceratodon	p.182
Pottiales	Pottiaceae	Barbula	p.184
		Syntrichia	p.186
Hedwigiales	Hedwigiaceae	Hedwigia	p.188
Splachnales	Splachnaceae	Splachnum	p.190
	Meesiaceae	Paludella	p.192
Bartramiales	Bartramiaceae	Bartramia	p.194
Bryales	Bryaceae	Bryum	p.196
		Rhodobryum	p.198
Orthotrichales	Orthotrichaceae	Zygodon	p.200
		Macromitrium	p.202
		Leiomitrium	p.203
Orthodontiales	Orthotrichaceae	Orthodontium	p.204
Rhizogoniales	Calomniaceae	Pyrrhobryum	p.206
Aulacomniales	Aulacomniaceae	Aulacomnium	p.208
Hypnodendrales	Racopilaceae	Racopilum	p.210
	Spiridentaceae	Spiridens	p.211
Ptychomniales	Ptychomniaceae	Ptychomnion	p.212
Hypopterygiales	Hypopterygiaceae	Hypopterygium	p.214
Hookeriales	Daltoniaceae	Daltonia	p.216
	Hookeriaceae	Hookeria	p.218
	Callicostaceae	Cyclodictyon	p.220
Hypnales	Hypnaceae	Hypnum	p.222
	Myuriaceae	Myurium	p.224
	Amblystegiaceae	Drepanocladus	p.226
	Fontinalaceae	Fontinalis	p.228
	Sematophyllaceae	Sematophyllum	p.230
	Orthostichellaceae	Orthostichella	p.231

OPPOSITE | *Dawsonia superba*, the world's tallest moss, in the Indonesian rainforest.

TAKAKIA

The Japanese name for *Takakia*, *nanjamonja-goke*, translates as "impossible moss"—a fitting name for a moss so puzzling that for more than 100 years it was classified as a liverwort. Unlike other mosses, the leaves of *Takakia* are deeply forked into two to four segments. For a long time, only *Takakia* plants without sporophytes were known, and based on leaf morphology these were originally placed in the liverwort genus *Lepidozia*. In 1990 *Takakia* plants with sporophytes were discovered, and as these were clearly aligned with the mosses, *Takakia* was reclassified. Among its many unique features are the segmented leaves, a root-like branching system, and capsules that open by a spiral split along the wall.

Takakia belongs to an ancient lineage that diverged from the mosses an estimated 390 MYA. Fossil evidence suggests that the two extant *Takakia* species have retained their current morphology for at least the last 165 MYA. Populations of *Takakia* on the Tibetan Plateau predate the uplift of the Himalayas that began 65 MYA, raising these plants to their current habitat over 13,000 ft (4,000 m) in altitude. *Takakia* has needed to adapt rapidly to survive in these extreme conditions and is extremely tolerant of both UV radiation and freezing conditions. However, *Takakia* is facing its greatest challenge yet. Global warming is increasing at such an unprecedented rate that *Takakia* cannot adapt quickly enough. Its populations are decreasing, and after nearly 400 million years, this unique lineage of plants may finally meet its match as a victim of human-induced climate change.

LEFT | The upright rigid shoots of *Takakia ceratophylla* can be seen here, along with the distinctive forked leaves of the genus.

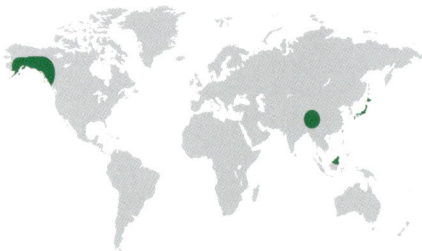

DISTRIBUTION
Western North America, Central and Eastern Asia

ETYMOLOGY
After Noriwo Takaki (1915–2006), who discovered *Takakia lepidozioides*

NUMBER OF SPECIES
2 accepted species

APPEARANCE
Bright-green turfs with shoots less than ⅜ in (1 cm) tall and a creeping root-like system below. Leaves in 2–4 segments, cells with numerous small oil bodies. Dioicous. Sporophytes very rare, capsule dehiscing along a single spirally suture line

ABOVE | A luxuriant patch of *Takakia lepidozioides* growing on wet rocks by a waterfall in British Columbia.

RIGHT | *Takakia lepidozioides* produces abundant capsules, which can be seen here with their distinctive spiral openings.

HABITAT
In cool, temperate to Arctic alpine regions, over moist soils, and rocky crevices in humid sites

SPHAGNUM

Sphagnum is a unique group of mosses that diverged from the moss lineage very early on in their evolutionary past. The structure of *Sphagnum* mosses is unique among bryophytes. They have small, well-spaced leaves on the main stems. The branches coming off this main stem are bunched together and comprise slender pendent branches that droop down covering the stem and thicker spreading branches. *Sphagnum* leaves are one cell thick and contain two types of cell. Most of the leaf consists of large empty dead "hyaline" cells whose function is to store water. These have gaps in their walls and usually have bands of spiral thickenings called fibrils. Alternating between these is a network of small green living cells. This unique cell structure allows the plants to absorb water quickly and release it very slowly. These mossy sponges can absorb up to 20 times their own weight in liquid.

The capsules are raised on a short stalk of gametophyte tissue called a pseudopodium that elongates when the capsules are mature. The dark, rounded capsules lack a peristome. Pressure inside the capsule increases as it dries out, and when

DISTRIBUTION
Widespread in temperate, montane, and boreal regions

ETYMOLOGY
Greek *sphágnos* = an ancient term for "moss" or "lichen"

NUMBER OF SPECIES
300 accepted species

APPEARANCE
Large plants, green, pink, or red. Stems with branches clustered into fascicles. Branch leaves egg-shaped to narrow, lacking costa, large hyaline cells with spiral fibrils and smaller green chlorophyllose cells. Mostly dioicous. Sporophyte with pseudopodium elevating the round capsule, which lacks a peristome

HABITAT
Found in wet sites such as nutrient-poor wetlands or wet woodlands

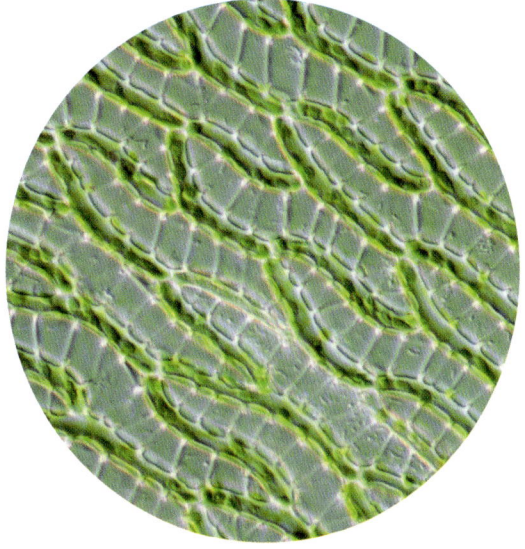

OPPOSITE LEFT | The golden-orange leaves of *Sphagnum lindbergii* are arranged in five ranks along the stem, giving the plant a particularly beautiful appearance.

OPPOSITE RIGHT | Unusually for a *Sphagnum*, this crimson *S. warnstorfii* is tolerant of habitats high in calcium, whereas most *Sphagnum* species prefer acidic conditions.

ABOVE LEFT | *Sphagnum subnitens* with abundant sporophytes, demonstrating the round capsules elevated on rather ephemeral-looking pseudopodia.

BELOW RIGHT | A single shoot of *Sphagnum russowii* showing the dome-like capitulum "head" and the branches down the stem arranged into bundles.

ABOVE RIGHT | Microscopic view of the *Sphagnum* leaf showing a continuous network of live green cells surrounded by numerous wide, partitioned chambers shown as pale gray in this photo.

mature the lid is shot off in a mini explosion, dispersing the spores.

Recent research has found that *Sphagnum* growth is influenced by moonlight. The surface of the bog is maintained as a fairly even surface, which helps reduce water loss by evaporation. New findings suggest that the levels of moonlight trigger the plants into a synchronized growth cycle, accelerating around a new moon and slowing down around a full moon.

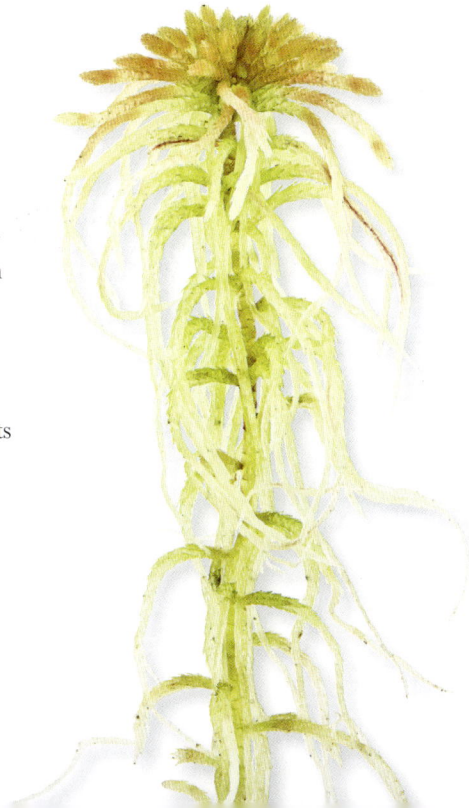

ANDREAEA

BELOW | A close-up view
of *Andreaea rupestris* shoots
highlights the reddish-orange
hues of this species and the
leaves that lack a costa.

The dense cushions of dark-red or almost black mosses on bare rock surfaces quickly become recognizable as *Andreaea* when hiking in the mountains. They cling tightly to the surface with rhizoids that penetrate small cracks in the rock. They are often quite brittle plants, with fragments of leaves breaking off easily when touched. On closer inspection the capsules of these mosses are especially distinctive. Rather than releasing spores from the top of the capsule via a lid, as with most mosses, the capsules are split lengthways into four to ten lobes joined at the top. A central column runs from the base of the capsule to the top, and as the capsule dries out, this column shrinks, pulling the capsule top down and resulting in the segments bulging outward, rather like a paper lantern. The spores can then disperse through the wide slits that open up.

Andreaea has such a unique morphology that it is placed in its own class, the Andreaeopsida. The capsules are raised on a stalk-like structure that looks like a seta but is in fact formed from tissue of the leafy gametophyte and is not part of the sporophyte. This haploid "pseudopodium" is a structure found only in the Andreaeales and the Sphagnales. The protonema in *Andreaea* is thalloid rather than filamentous as it is in most mosses. *Andreaea* are

DISTRIBUTION
Widespread, mostly temperate to
tropical montane

ETYMOLOGY
After Johann Gerhard Reinhard
Andreä (1724–1793), chemist and
natural scientist of Hannover

NUMBER OF SPECIES
87 accepted species

APPEARANCE
Small to medium-sized plants, stems upright,
red-black, forming cushions or mats. Leaves
with costa single, branched, or absent, cells
thick-walled, papillose, or smooth. Mostly
monoicous. Capsules elliptical, elevated by
pseudopodium, spores dispersed through
longitudinal valves. Calyptra tiny

ABOVE | *Andreaea* species tend to form dark reddish-purple cushions that cling tightly to the rocks, as show with this *A. blyttii*.

BELOW | The characteristic capsules of *Andreaea* open with longitudinal valves, giving them a lantern-like appearance.

usually monoicous, meaning one plant produces both male and female reproductive organs. Such close proximity of the sexes helps ensure a higher rate of successful fertilization.

HABITAT
On acidic rocks such as granite, in cool temperate to tropical montane regions

ANDREAEOBRYUM

BELOW | Dark cushions
of *Andreaeobryum* cling
to the surface of boulders,
photographed here in Far
East Russia.

The dark, almost black mats of *Andreaeobryum* adhering tightly to rocks look rather like *Andreaea*. The two genera even share the distinctive capsules with longitudinal opening valves. However, there are on more detailed inspection some key differences that start

DISTRIBUTION
Restricted to Arctic and sub-Arctic areas of northwest North America

ETYMOLOGY
The genus *Andreaea* + Latin *bryon* = "moss," referring to a resemblance to the *Andreaea* mosses

NUMBER OF SPECIES
Only 1 species

APPEARANCE
Stems upright, ¾–1½ in (2–4 cm) tall, dark green, red to black forming mats. Leaves ovate or lanceolate, strongly curved, with indistinct costa, cells smooth. Occasionally with flagelliform branches with tiny appressed leaves. Dioicous. Setae short. Capsules angular and conical, dehiscing via 4–8 longitudinal valves. Calyptra large, covering whole capsule

HABITAT
On calcareous rocks

to suggest that these two groups are less closely related than may at first appear. In *Andreaeobryum* the capsule shape is angular whereas in *Andreaea* it is elliptical and the top of the capsule erodes when mature so the valves become free, not attached at the top. The calyptra is also different in *Andreaeobryum* since it is much larger and covers the capsule. Developmentally, the stalk that elevates the capsule is a true seta in *Andreaeobryum* since it is derived from the sporophyte tissue, not an extension from the leafy gametophyte. *Andreaeobryum* is dioicous, meaning male and female reproductive structures occur on separate plants, unlike *Andreaea*, which is usually monoicous. *Andreaeobryum* is also distinguished by its habitat since it only grows on limestone rocks whereas *Andreaea* is found on more acidic rock types such as granite.

There is only the one species represented by the Andreaeobryales order, *Andreaeobryum macrosporum*, which was only discovered relatively recently in the 1970s. It produces much larger spores than any of the *Andreaea* species, hence the name. It is a rare plant within its restricted range of northwestern America and far-eastern Russia. Where it does occur it may grow in some abundance and can form a striking contrast of black moss cushions against the white limestone.

ABOVE | A close view of an *Andreaeobryum* cushion showing the curved leaves and large calptrae that are dark above and pale below.

123

OEDIPODIUM

BELOW | A single leaf of *Oedipodium griffithianum*, rounded in the upper part and narrowing to a slim base.

The succulent green shoots of *Oedipodium* have broad, rounded leaves that get more crowded further up the stem, where they become clustered into rosettes. Examination of an individual leaf would reveal the narrow base has long delicate hairs extending from the margin. The most striking feature of this plant are the chunky pale-green sporophytes, which have a long, thickened capsule neck. There are usually clusters of pale gemmae produced on short stalks from the leaf axils.

The Oedipodiales includes only one species: *Oedipodium griffithianum*. The relationship of *Oedipodium* to other mosses has been the subject of much past debate, and the genus has previously been placed within the Funariales or the Splachnales. This ambiguity in classification has often been the case for moss taxa that lack the informative taxonomic characters of a peristome. DNA sequence data has confirmed a more isolated evolutionary position for *Oedipodium* among the lineages of early diverging mosses. It is suggested to be closely related to the Tetraphidales.

Oedipodium has a scattered and rather disjunct distribution in cool temperate regions of the world. This wonderful and bizarre little plant is threatened with extinction across much of its range. The reason for its rarity is unclear since it does not appear to

DISTRIBUTION
Northern North America and southern South America, Europe, far-eastern Asia

ETYMOLOGY
Greek *oidema* = "swelling" + *podium* = "platform," referring to the swollen capsule neck

NUMBER OF SPECIES
Only 1 known species

APPEARANCE
Pale green to ⅜ in (1 cm) tall. Leaves 3/16–⅛ in (2–3 mm) long, rounded above, narrowing with hairs at base, costa single, laminal cells large, hexagonal, smooth. Monoicous. Seta green and fleshy. Capsule with swollen neck, peristome absent. Calyptrae hooded. Asexual reproduction by green gemmae

ABOVE | *Oedipodium griffithianum* plants showing the large gemmae attached to the upper surface of the leaves.

RIGHT | The thick seta and swollen capsule neck are distinctive of *Oedipodium*.

HABITAT
On moist soil and rocky crevices, often in alpine regions

TETRAPHIS

BELOW | These modified leafy cups contain the plant's vegetative propagules, the gemmae ready to splash out when raindrops land in the cups and grow into new plants.

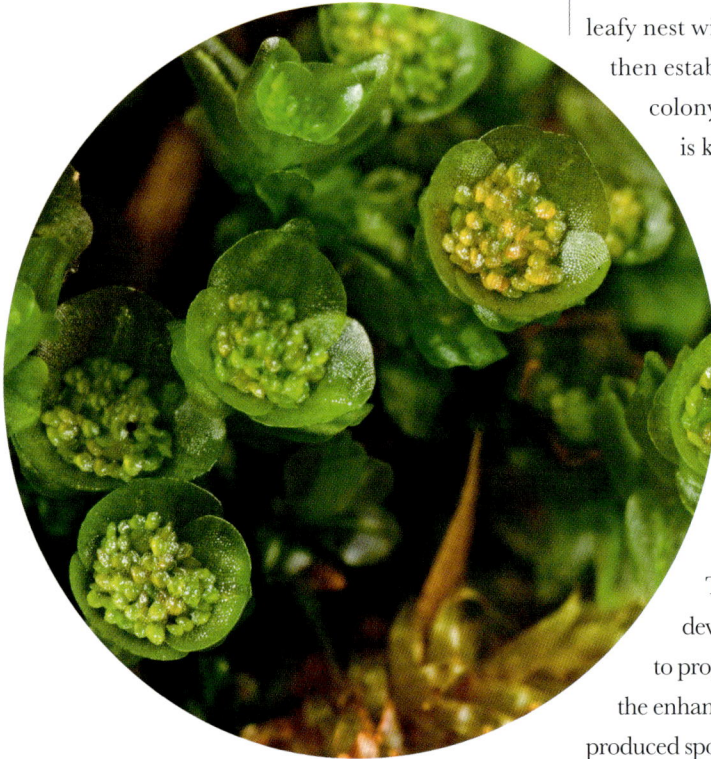

*T*etraphis is most typically found carpeting rotting logs and stumps in temperate woodland. The tips of some of its vegetative shoots may end in a rosette of modified leaves within which lie a cluster of disc-shaped green gemmae, like eggs in a tiny emerald-green nest. Each gemma is a vegetative bud that can grow into a new leafy plant. A raindrop landing in the leafy nest will splash out the gemmae, which can then establish into new plants, extending the colony whenever a chunk of rotting wood is knocked to the floor.

There will come a time when the wooden stump is fully occupied, and this triggers a change in the moss's reproductive strategy. It now switches from making gemmae to producing female shoots. When fruiting, *Tetraphis* is unmistakable since each long cylindrical capsule ends in four large peristome teeth, which is very unusual among mosses. The abundant capsules that now develop disperse spores enabling the moss to propagate much further afield. As well as the enhanced dispersal potential, sexually produced spores will have a greater genetic diversity,

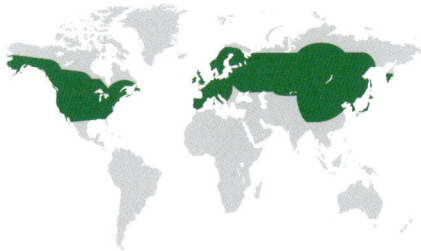

DISTRIBUTION
North America, Europe, far-eastern Russia, China, Japan

ETYMOLOGY
Greek *tetra* = "four" + *fissus* = "split," referring to the 4 peristome teeth

NUMBER OF SPECIES
2 accepted species

APPEARANCE
Upright shoots less than ⅝ in (1.5 cm) high. Leaves egg-shaped with single costa. Monoicous. Capsules with distinctive 4-toothed peristome. Vegetative shoots often ending in rosette of leafy bracts holding gemmae

more suitable for establishing in new frontiers. In contrast, the vegetative buds of the gemmae produce plants that are clones of the parent, well suited for thriving in the same micro-habitat. The intricate life strategy of *Tetraphis* was studied and documented in detail by North American bryologist Robin Wall Kimmerer, who referred to *Tetraphis* as a "sequential hermaphrodite," with the ability to change its gender from male to female as the colony gets crowded.

HABITAT
On well-decayed wood, may also occur on rocks such as sandstone if very moist

ABOVE | Plants of *Tetraphis pellucida* can typically be found on rotting wood and tree stumps in woodland.

LEFT | *Tetraphis* gets its name from these distinctive capsules with four large peristome teeth.

DAWSONIA

The Polytrichales are an order of giants and have a number of sophisticated adaptations that challenge the misconception of bryophytes as small and simple plants!

The long, narrow leaves have a broad costa that fills most of the leaf. The surface of the costa has rows of cells called lamellae running lengthways down the center of the leaf like green ribbons standing on edge. These lamellae increase the surface area for photosynthesis and provide space for gas exchange since the tips of the lamellae have a waxy coating that repels water. The stems of most Polytrichales have a well-developed conducting system comprising narrow cells in their stems called hydroids, which conduct water, and leptoids, which transport nutrients.

Dawsonia superba has the accolade of tallest moss in the world (see page 115), and at first glance could be mistaken for a clump of pine tree seedlings. *Dawsonia* has a very unusual peristome consisting of a brush-like bundle of hairs, which differentiates it from other members of the order.

LEFT | A mature sporophyte capsule showing the unique hair-like peristome of *Dawsonia*.

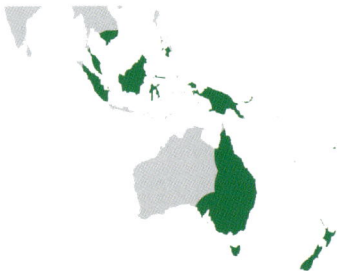

DISTRIBUTION
Malaysia, New Guinea, Australia, New Zealand

ETYMOLOGY
After Dawson Turner (1775–1858), an English banker, botanist, and antiquarian

NUMBER OF SPECIES
10 accepted species

APPEARANCE
Plants reaching up to 27½ in (70 cm) tall. Leaves narrow, up to 1⅜ in (3.5 cm) long, with sheathing base appressed to stem and spreading limb, costa broad with numerous lamellae. Dioicous. Capsule with 2 sharp angles. Peristome formed of rings of bristle-like bundles of cells. Calyptra hairy

HABITAT
On soil, in moist forests

POLYTRICHUM

The genus *Polytrichum* is often described as the "Haircap" mosses, referring to the densely hairy calyptra. They are distinguished from other genera in the Polytrichales by the combination of the leaves with a sheathing base that clasps the stem and some particular details of the sporophyte. The capsules have four prominent angles, making them rather box-like. The peristome of the Polytrichales is nematodontous. In *Polytrichum* the teeth are joined at their tips to a membrane called the epiphragm that covers the capsule mouth. Spores disperse through small openings between the teeth.

Most *Polytrichum* species produce separate male and female plants. The male sex organs or antheridia are produced in a nest of differentiated leaves called the perigonia at the tips of shoots. These function as splash cups to catch rain and cascade the sperm, ideally to a nearby female plant.

Polytrichum can grow in such extensive large hummocks that in some cultures it has been traditionally used as a broom for sweeping.

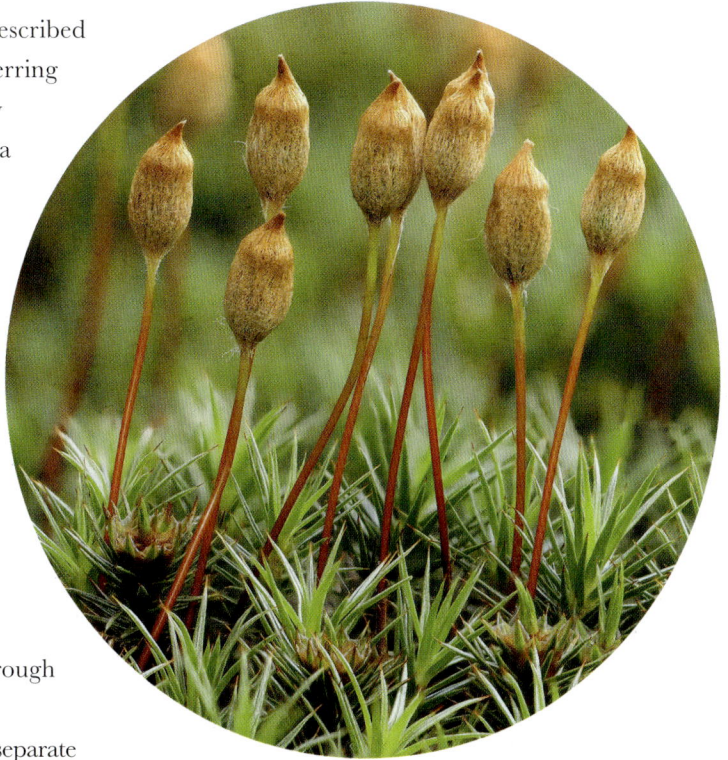

ABOVE | *Polytrichum juniperinum* is a widespread species, photographed here in England but found across temperate and tropical montane regions of the world.

DISTRIBUTION
Widespread

ETYMOLOGY
Greek *polús* = "many" + *trichós* = "hair," referring to the hairy calyptra

NUMBER OF SPECIES
40 accepted species

APPEARANCE
Dark-green plants, usually large, up to 4 in (10 cm) tall, in extremes to 20 in (50 cm). Leaves narrow and sharply pointed, with sheathing base appressed to stem, costa broad with numerous lamellae. Dioicous. Capsule 4-angled. Peristome of 64 teeth attached to oval epiphragm. Calyptra hairy

HABITAT
On soil, in moist areas of woodlands or sheltered banks

BUXBAUMIA

The *Buxbaumia* genus is only noticeable when sporophytes are present, since the leafy gametophyte phase is so reduced it can barely be seen. The capsules, however, are enormous by bryophyte standards and earn some species the common nickname of "Bug-on-a-Stick" moss.

Buxbaumia grows from a persistent protonema that extends over suitable habitat. Tiny buds are produced from the tips of protonemal branches and these develop into a reduced gametophyte bearing the male or female reproductive organs. The sporophyte is initially green with a robust seta that has a small central conducting strand. The capsules are very asymmetric, with one side usually forming a flattened surface. It has been suggested that the pressure of raindrops falling on this surface compresses the air within the capsule, causing the spores to be puffed out through the narrow mouth.

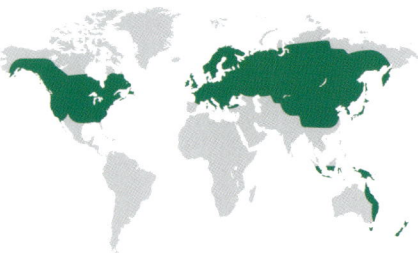

DISTRIBUTION
Widely distributed in northern hemisphere, also found in Australia and New Zealand

ETYMOLOGY
After Johann Christian Buxbaum (1693–1730), a German physician and botanist

NUMBER OF SPECIES
11 accepted species

APPEARANCE
Protonema persistent forming brownish mats. Gametophyte phase of plant highly reduced. Dioicous. Seta ⅛–⁷⁄₁₆ in (3–11 mm) tall, roughened. Capsule strongly asymmetric with upper surface flattened. Peristome exostome of short teeth and endostome a pleated membrane. Calyptra small

OPPOSITE LEFT | The upper surface of *Buxbaumia aphylla*'s brown capsules is flattened, with angular edges.

OPPOSITE RIGHT | The large green upright capsules of *Buxbaumia viridis* are unmistakable.

ABOVE | *Buxbaumia viridis* grows on dead wood such as rotting logs and tree stumps in forests.

HABITAT
Usually associated with rotting wood in forests

Buxbaumia mosses have a particularly complicated peristome structure, which is derived from many concentric cell layers.

These are rare mosses throughout their range. Some species are considered to be threatened and species in Europe have been found to be in decline, though North American populations seem more stable. *B. viridis* has been reported with 25–50 percent declines in its European populations and has been offered legal protection in Europe as part of a conservation strategy. These measures have helped to increase the resources available for recording the detailed distribution of this species and studying its ecology. Causes of its decline are thought to relate to a decrease in available rotting dead wood left in forests and reduced humidity due to the removal of trees or increased frequency of droughts.

DIPHYSCIUM

Female plants of *Diphyscium* are immediately recognizable when their distinctive capsules are present. These are large, ovoid, and asymmetric, immersed in a nest of long-awned perichaetial leaves. The leafy shoots of the female plants are very much reduced, but the male plants can look confusingly similar to mosses in the

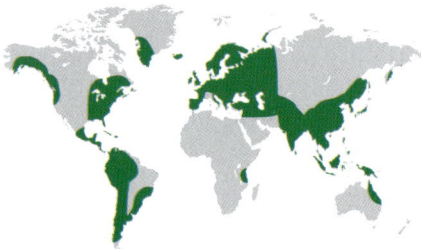

BELOW | The capsules of *Diphyscium foliosum* sit within a crown of long, narrow leaves with toothed margins.

DISTRIBUTION
Widespread, though in Africa only recorded from Tanzania

ETYMOLOGY
Greek *di* = "two" + *physce* = "bladder," referring to the space between the capsule wall and spore sac

NUMBER OF SPECIES
17 accepted species

APPEARANCE
Short turfs up to ³⁄₁₆ in (5 mm) tall. Leaves tongue-shaped, costa single, cells smooth or papillose. Dioicous. Perichaetial leaves strongly differentiated with very long awns. Seta very short. Capsule large, ¹⁄₁₆–¹⁄₈ in (2–4 mm), ovoid, asymmetrical. Peristome with exostome reduced or absent, endostome a pleated membrane. Calyptra small

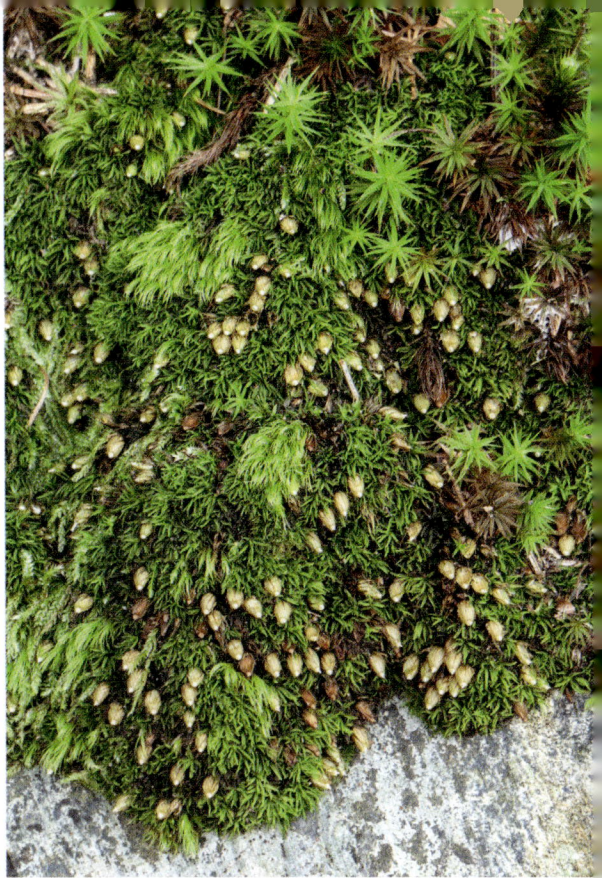

Pottiaceae. Their leaves are long and narrow with a rounded tip. Clues that these in fact belong to *Diphyscium* come from the leaves looking dull and opaque, since the leaf is two or three cells thick, and also the costa is quite faint while in the Pottiaceae it is usually very noticeable. The capsules have a narrow mouth from which the peristome protrudes as a white, twisted cone, which is the pleated membrane. The capsule shape and peristome structure are similar to *Buxbaumia*, and the two orders Buxbaumiales and Diphysciales are thought to be closely related.

The charismatic capsules of *Diphyscium* have led to several common names being used for this moss, such as "Nut" moss or "Grain-of-Wheat" moss. In North America they are also known as "Powder Gun" moss, alluding to the spore dispersal strategy employed by *Diphyscium*. The capsules are held close to the ground, unlike most mosses, which elevate them on a long seta to catch the wind above the atmospheric boundary layer. In an approach shared with *Buxbaumia*, they rely on raindrops to land on the broad surface of the capsules and eject a cloud of spores.

HABITAT
On shaded soil banks and forest floors

TOP | The large brown capsules of *Diphyscium foliosum* stand out among cushions of other green mosses.

ABOVE | *Diphyscium mucronifolium* has a disjunct distribution, with small populations around the eastern USA, where this photo was taken, and also in eastern Asia.

133

GIGASPERMUM

BELOW | Tiny plants of
Gigaspermum repens forming
a silvery turf over soil in the
Western Cape, South Africa.

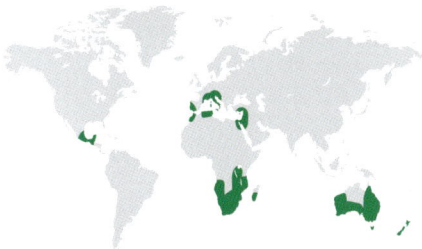

Although the individual plants of *Gigaspermum* are tiny, they can form an eye-catching silvery turf over bare soil. The upright leafy shoots arise from leafless underground root-like branches. The spores are produced in early spring, after which the shoots die back to their drought-resistant underground branch system until the shoots regrow the following year.

DISTRIBUTION
Predominantly found in the southern hemisphere but also Mexico and Mediterranean Europe

ETYMOLOGY
Greek *gigas* = "giant" + *sperma* = "seed," referring the relatively enormous capsule and/or spores

NUMBER OF SPECIES
2 accepted species

APPEARANCE
Tiny green, occasionally silvery plants up to ³⁄₁₆ in (5 mm tall), shoots arising from a fleshy subterranean rhizome. Leaves teardrop-shaped with pointed end at insertion, costa single or absent. Monoicous. Seta very short. Capsule immersed, lacking peristome. Calyptra minute

Leaves on sterile shoots are rounded and spread away from the stem. On fertile shoots the leaves are larger, whitish, and narrow, to a slender acuminate point. The relatively enormous capsules sit within a cluster of leaves. These larger perichaetial leaves protect the developing sporophyte and then turn pale and papery when mature, usually with a long hairpoint. The capsule mouth is very wide, lacking a peristome, and the spores inside are huge.

The plant we now know as *G. repens* was first scientifically described as a new species by the nineteenth-century botanist Ferdinand von Mueller based on a specimen collected in New South Wales, Australia. In something of a botanical faux pas, however, Mueller described this "moss-like plant" as a flowering plant under the genus *Trianthema* in the Aizoaceae family! The error occurred because Mueller was thrown off by the unusually large capsules with huge spores visible inside, which he mistook for seeds.

In some regions of Australia *Gigaspermum* used to be a common plant carpeting roadsides. It has now sadly declined since it is being out-competed by weeds whose robust growth is fueled by nitrate fertilizer pollution.

ABOVE | The large capsule lid of this *Gigaspermum repens* capsule is on the verge of falling away, revealing the very large spores inside.

BELOW | The minute pointed tips of the leaves can give *Gigaspermum repens* a spiky appearance.

HABITAT
On soil, typical dry, bare places

DISCELIUM

There is only one species in this order: *Discelium nudum*. It is a plant that is only noticed when it is producing its towering sporophytes. Rather like *Buxbaumia*, *Discelium* produces these sporophytes from a very small, reduced leafy gametophyte that develops from a persistent green protonema. The tall setae raise the capsules up to heights of around ³⁄₈ in (1 cm) but in extreme examples can reach up to 1 ¼ in (3 cm). The capsules are slightly elongate and angled horizontally at the top of the seta, with 16 prominent reddish-brown peristome teeth.

Discelium is a pioneer of precarious habitats such as exposed reservoir mud or clay banks. Despite its wide global range, with a circumboreal distribution, it is a rare plant of scattered occurrence. It appears to thrive on instability, as it is easily outcompeted by other bryophytes. Its protonema produces colorless tubers that are concentrated in a layer approximately ³⁄₈ in (1 cm) below the surface of the clay, attached to a network of underground rhizoids. Research into the purpose of these tubers found them to be short-lived but quick to germinate. Over winter, repeated frost action and running water cause the surface of these clay banks to break up and wash away, exposing the rapidly germinating tubers in the spring. This reproductive strategy gives *Discelium* a competitive advantage in colonizing the clay banks

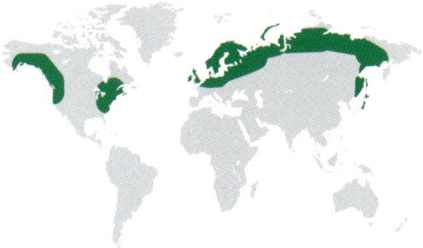

DISTRIBUTION
Widespread across northern hemisphere

ETYMOLOGY
Greek *di* = "two" + *skélos* = "legs," referring to peristome teeth that often have longitudinal perforations

NUMBER OF SPECIES
Only 1 species accepted

APPEARANCE
A persistent protonema giving rise to tiny leafy plants, up to ¹⁄₁₆ in (2 mm) high. Leaves brownish, ovate-lanceolate. Dioicous. Seta very tall, ⅛–1¼ in (4–30 mm), twisted when dry. Capsule elongated. Peristome orange-red, with 16 exostome teeth that are striated, often with longitudinal gaps and endostome membrane. Rhizoidal tubers present

LEFT | The oval-shaped capsules of *Discelium* are held at an angle atop a seta that can reach heights of up to 10 in (25 cm).

BELOW | The sporophytes of *Discelium* appear is if they are arising directly from the bare soil, and the setae can be characteristically undulate.

before other bryophytes have a chance to become established. The plants also produce a huge volume of spores that can disperse widely and are longer lived, allowing them to colonize new suitable habitats.

OPPOSITE | The highly reduced leafy shoot is just visible here at the base of the towering seta.

HABITAT
Moist clay or silty banks, occasionally exposed reservoir margins

ENCALYPTA

The large pale conical calyptrae that completely cover the capsules are very distinctive for *Encalypta* and have resulted in its common name in some regions of "Extinguisher" moss.

The plants form dense tufts typically found growing on thin soil over rocky outcrops in exposed dry sites, where conserving and making efficient use of available water is of utmost importance. The crowded leaves look translucent when moist and spread out from the stem. When dry, however, the leaves become contorted and twisted—a strategy to reduce water loss from exposed leaf surfaces. Under the microscope we can see that the cells in the upper part of the leaf are partially obscured by numerous tiny, branched papillae. Conduction of water in bryophytes is predominantly external. These papillae covering the leaf surface form a network of interconnecting capillary spaces that allow water to move across the leaf by capillary action. The cells at the leaf base of *Encalypta* are large, rectangular, and empty with perforations in the cells' walls, and they likely perform a role in water storage.

LEFT | The long, cylindrical calyptrae of *Encalypta ciliata* are frilled at the base.

OPPOSITE TOP | A species typically associated with montane habitats, *Encalypta rhaptocarpa* is found here perched on the alpine peaks of Switzerland.

DISTRIBUTION
Widespread

ETYMOLOGY
Greek en- = "within" + *kalyptō* = "to conceal," alluding to the large calyptra that covers the capsule

NUMBER OF SPECIES
41 accepted species

APPEARANCE
Plants small to medium, 3/16–1 1/4 in (5–30 mm) high. Leaves 1/16–1/8 in (2–4 mm) long, tongue-shaped with apex broadly rounded, occasionally lanceolate, costa single, cells ornamented with large, branched papillae. Monoicous. Seta elongate, often quite short. Capsule long, cylindrical, peristome single, double or absent. Calyptra large, sometimes with fringe below

HABITAT
Growing on soil or rocks in dry habitats

The diversity of peristome types within *Encalypta* is unusual and has made placing this genus into classifications that traditionally use peristome features as a way to categorize different groups of mosses quite challenging. Molecular DNA evidence has helped to confirm that the Encalyptales are most closely related to the Disceliales and Funariales.

RIGHT | When viewed microscopically, the upper leaf surface of the *Encalypta* mosses is seen to be covered in minuscule branched papillae.

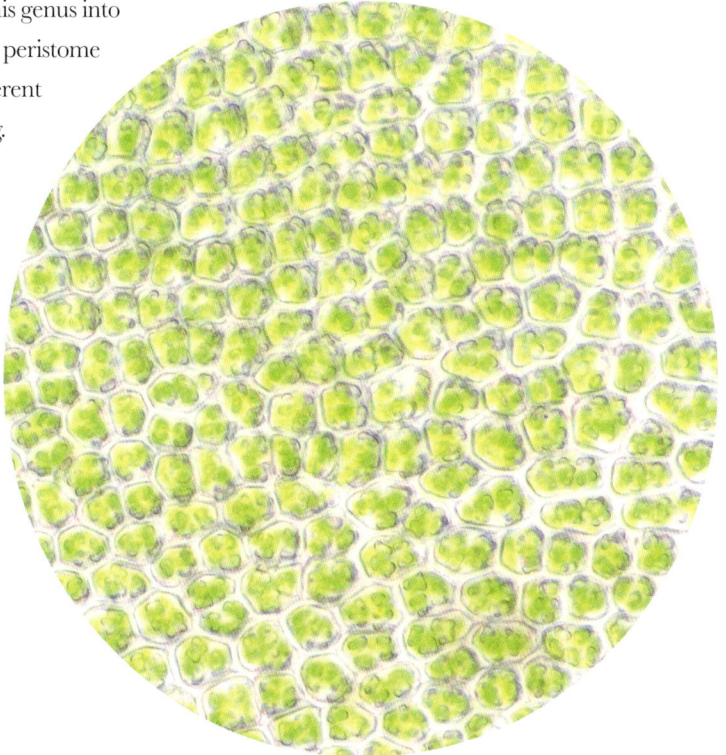

PHYSCOMITRIUM

LEFT | The large, thin-walled cells of the *Physcomitrium patens* leaf are visible with a hand lens.

The Funariales is an order of small mosses characterized by broad leaves with large cells and peristomes (when present), with the two rows of peristome teeth positioned opposite each other. The species mostly adopt an ephemeral life cycle on disturbed open soils. Identification usually relies on the sporophyte characters, and fortunately these are often present. Until recently *Physcomitrella* and *Aphanorhegma* were treated as separate genera in the Funariaceae, but these species have now been united under *Physcomitrium*, leading to a slightly broader concept for this genus in terms of morphology.

Since the 1960s, *Physcomitrium patens* has been used in research as a model species, meaning it has been studied intensively to investigate specific biological processes. An initial culture was grown from a single spore collected from a plant in Huntingdon, UK, in 1962, and this quickly became the standard laboratory strain passed to laboratories around the world. It is a convenient species to grow in culture and has a short generation cycle. It has been used to investigate a wide range of processes

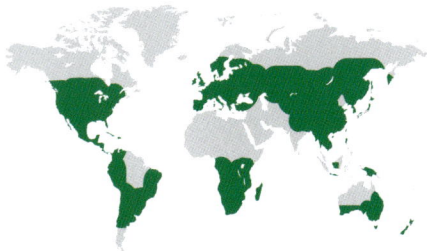

DISTRIBUTION
Widespread

ETYMOLOGY
Greek *physce* = "bladder" + *mitrion* = "little cap," referring to the shape of the operculum

NUMBER OF SPECIES
64 accepted species

APPEARANCE
Plants up to $1/16$–$3/8$ in (2–10 mm) tall. Leaves with entire or serrate margins, costa single. Monoicous. Seta short, up to $1/64$ in (0.2 mm), occasionally $5/8$ in (15 mm), tall. Capsules of some species lack opening lid, so spores disperse when capsule wall breaks down, peristome absent. Calyptra small or large with spreading lobes

HABITAT
Often on exposed soil, such as banks of waterbodies exposed by low waters

in evolutionary and developmental biology. A major attraction of *Physcomitrium* to evolutionary biologists is that it is especially amenable to precise genetic manipulation. Specific genes can be targeted and disrupted so that their function in the plant's biology can be inferred. *P. patens* was the first non-seed plant to have its whole genome sequenced. The extensive research around this moss continues to advance, with applications now being investigated in its role in the production of biopharmaceuticals.

BELOW | A mature, globose capsule of *Physcomitrium patens* containing fully developed spores turns a striking orange.

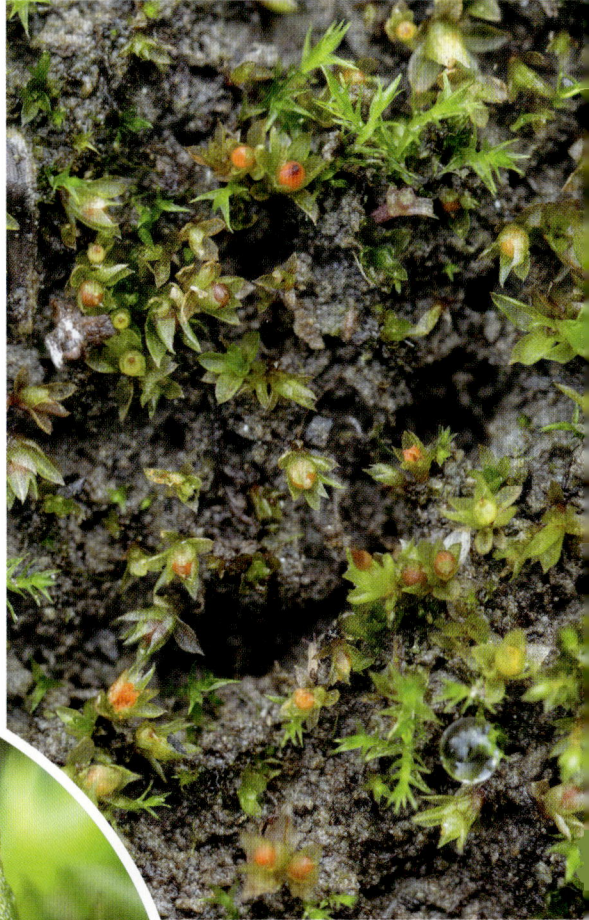

ABOVE | *Physcomitrium patens* is a colonizer of bare mud, such as at exposed reservoir margins, where it must complete its life cycle quickly over the drier period of the year before its habitat is under water again.

TIMMIA

*T*immia is best described as looking like a *Polytrichum* without the leaf lamellae. It has the long, toothed leaves that spread from sheathing leaf bases. The peristome structure in the Timmiales is unique, with an exostome of 16 large teeth and endostome of 64 filaments that arise from a basal membrane. Another

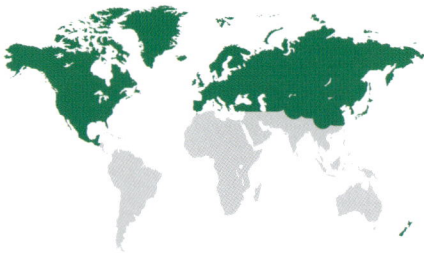

BELOW | *Timmia austriaca* could be confused for a small *Polytrichum* but its leaves are thicker and more opaque.

DISTRIBUTION
Widespread across northern hemisphere and also New Zealand and Pacific islands

ETYMOLOGY
After Joachim Christian Timm (1734–1805), German apothecary, botanist, and town mayor

NUMBER OF SPECIES
8 accepted species

APPEARANCE
Plants up to 4¾ in (12 cm) tall. Leaves lanceolate with sheathing base and spreading upper section, margins serrate, costa single. Dioicous or monoicous. Seta elongate up to 1⅜ in (3.5 cm). Capsule inclined or pendulous. Peristome with 16 large exostome teeth, and endostome is a pleated membrane. Calyptra hood-shaped

HABITAT
On soil or rocks

useful character that helps define *Timmia* is the shape of the upper leaf cells that bulge outward, giving the leaf surface a microscopically bumpy texture. It is a small, taxonomically isolated order containing just one genus, *Timmia*.

T. megapolitana is historically known by the common name "Indian Feather" moss in North America. The calyptra remains clasping the seta after falling away from the capsule. The capsules are inclined, and this persistent calyptra can point directly upward from the capsule base, resembling the lone feather of a headdress as traditionally worn by some Indigenous American people. *T. megapolitana* is widespread and extends into temperate regions, unlike the other species of *Timmia*, which are generally restricted to Arctic or mountainous sites. Species of *Timmia* are restricted to the northern hemisphere except for *T. norvegica*, which has a curiously disjunct distribution and also occurs in New Zealand. There it is only known from a few sites, and only sterile plants have ever been found.

ABOVE | Bright-green shoots of *Timmia bavarica*, a predominantly alpine-montane species.

LEFT | A single leaf of *Timmia austriaca* showing a strong costa, toothed leaf margins, and a sheathing base that is distinctly orange in this species.

CATOSCOPIUM

BELOW | The shiny, black capsules of *Catoscopium nigritum* on red setae.

The dense blackish cushions of *Catoscopium* have characteristically long dark-red setae with horizontally inclined shiny black capsules atop that resemble tiny golf clubs. There is only one species in the Catoscopiales: *C. nigritum*. This order is thought to represent the earliest divergence in the Dicranidae class of mosses, which are defined by having a peristome composed of a single ring of peristome teeth.

C. nigritum is a calcicolous species that grows on sandy ground by the sea, in rich fens, and in upland calcareous flushes. It requires a nutrient-poor habitat where there is little competition from larger plants. It also likes to be flushed by water that deposits calcium carbonate. This is a great example of how underlying geology can prescribe the remit of suitable habitat available for a fussy species. Although generally quite rare across its range, it can form extensive carpets in the right conditions. In some countries across Europe, however, it is at risk of extinction. This is due to pressures on specific habitats, such as coastal dune systems and lowland fens, which are threatened by land development and over-

DISTRIBUTION
Circumpolar across the northern hemisphere

ETYMOLOGY
Greek *katá* = "downward" + *skopéō* = "to look," in reference to the down-turned capsules

NUMBER OF SPECIES
Only 1 accepted species

APPEARANCE
Plants dark green to almost black, up to 2⅜ in (6 cm) tall. Leaves lanceolate with narrowly recurved margins. Dioicous. Seta elongate, 1⅓ in (8–24 mm) tall, red-black. Capsule inclined, glossy, curved with mouth pointing downward, peristome with 16 exostome teeth, endostome reduced. Calyptra hood-shaped

fertilization by pollutants that encourage the growth of weedy species that can out-compete smaller, less-competitive plants.

RIGHT | The leaves of *Catoscopium* are rather triangular in shape, with a strong costa.

ABOVE LEFT | A fertile cushion of *Catoscopium*; when the capsules mature they will darken to almost black.

RIGHT | *Catoscopium* can form deep cushions in optimal conditions.

HABITAT
On wet ground of calcareous habitats

DISTICHIUM

The pale-green silky cushions of *Distichium* with their long narrow leaves could be confused with several other genera, such as *Dicranella*, at first glance. Inspection of an individual shoot, however, shows that the leaves of *Distichum* are arranged in two rows on opposite sides of the stem, and this flattened appearance is distinctive for the genus. Each leaf clasps the stem with its broad sheathing base and then abruptly narrows into a long spreading limb.

The genus *Distichium* was previously considered part of the family Ditrichaceae. However, analysis of DNA data found Ditrichaceae to include genera that do not share a common ancestor. *Distichium* is one of the early branching lineages of the Dicranidae class of mosses, which are characterized by a peristome with a single ring of teeth. It is the only genus in the order Distichiales that was established in 2023 to reflect the evolutionary distinctiveness of this group of mosses.

LEFT | *Distichium capillaceum* clings to a rock face.

DISTRIBUTION
Widespread

ETYMOLOGY
Greek *dístichos* = "arranged in two rows," referring to the arrangement of the leaves

NUMBER OF SPECIES
11 accepted species

APPEARANCE
Slender plants forming dense tufts, yellow-dark green, ⅛–2⅜ in (1–6 cm) tall. Leaves distichous, lower part of leaf whitish, sheathing the stem, costa single. Monoicous. Seta elongate, yellow-red. Capsule cylindrical, peristome a single ring of 16 teeth strongly perforated or irregularly divided. Calyptra hooded

ABOVE | This close-up of a shoot of *Distichium capillaceum* shows the leaves in two ranks up the stem.

RIGHT | *Distichium capillaceum* has erect, cylindrical capsules whereas some other common members of *Distichium* hold their capsules at an angle.

HABITAT
On moist soil or rock in calcareous habitats

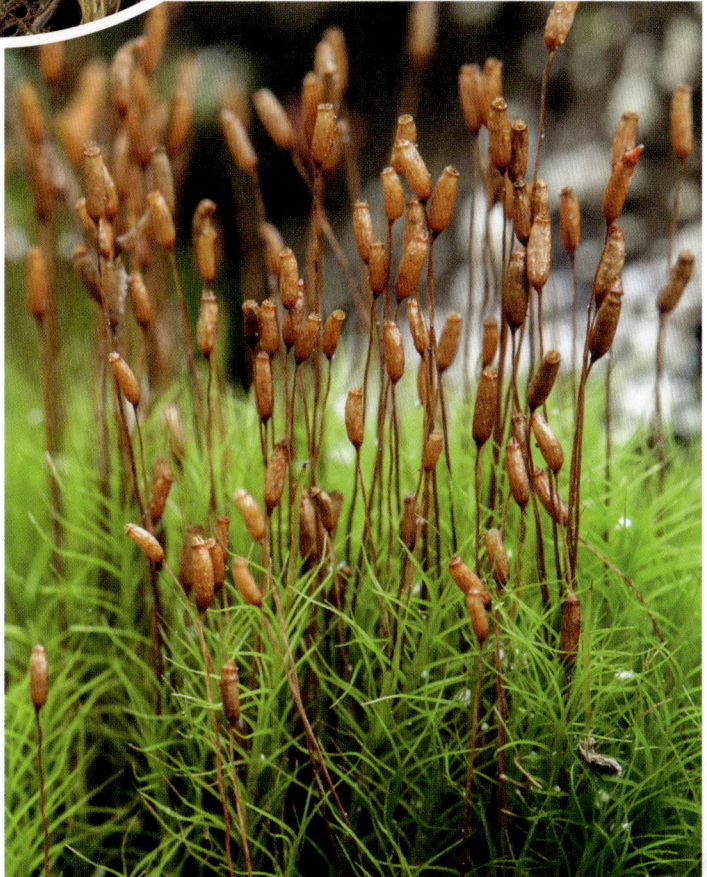

147

FLEXITRICHUM

Until recently the two species now classified as *Flexitrichum* were treated as part of the genus *Ditrichum*, but they have been found to be sufficiently distinct from their closest relations to warrant a segregated status in their own order. *Flexitrichum* are rather large mosses. Their leaves have broad sheathing bases, with the spreading part of the limb narrowing into a long, fine point filled by the costa. The leaves are often swept slightly to one side. They usually have a dense coating of rhizoids on the lower stem, which can help distinguish them from species of *Ditrichum*.

The two species *F. flexicaule* and *F. gracile* are quite similar to each other and can be challenging to identify. In fact, they were only recognized as two distinct species in 1985. There is a lot of

DISTRIBUTION
Widespread, predominantly northern hemisphere but also northern South America, New Guinea, New Zealand

ETYMOLOGY
Flexitrichum is a portmanteau of the most well-known species of the genus, *Ditrichum flexicaule*

NUMBER OF SPECIES
2 accepted species

APPEARANCE
Dense tufts ⅜–2¾ in (1–7 cm) tall. Leaves flexuous when dry, ⅛–¼ in (4–7 mm) long with base sheathing stem, narrowly lanceolate, costa single. Dioicous. Seta red-brown, up to 1 in (2.5 cm) long. Capsule erect, cylindric, peristome a single ring of teeth

HABITAT
Dry calcareous habitats, sand dunes, grasslands

morphological variation in these plants, which in the past has been interpreted with additional species recognized. Some of the variation in microscopic characters that has caused so much past confusion is thought to be attributable to environmental conditions, with levels of shade and humidity influencing the development of the plants. These are species that love lime, and they grow in calcareous grasslands, sand dunes, and limestone crags.

OPPOSITE | *Flexitrichum gracile* growing on a limestone wall in northern England, accompanied by *Ctenidium molluscum* in the lower left of the image.

BELOW | The leafy shoots of *Flexitrichum flexicaule* are matted with dense brown rhizoids at the base.

ABOVE | Long silky green leaves of *Flexitrichum gracile* with some shoots of the curly leaved *Ctenidium* poking through.

PSEUDODITRICHUM

P lants now known as *Pseudoditrichum mirabile* were first collected in 1948 near Great Bear Lake in northern Canada. However, it was not until 1974 that it was described as the new genus *Pseudoditrichum* based only on this single collection. In 2015 another population of this exceedingly rare moss was reported from the Anabar Plateau of northern Siberia.

The leafy shoots resemble mosses in the Ditrichaceae family, and indeed this plant could easily be overlooked as a small *Ditrichum* in the field. However, *Pseudoditrichum* plants have a peristome with a double ring of teeth positioned opposite each other. This indicates a very different evolutionary position to *Ditrichum*, which has a single ring of peristome teeth. It has been previously classified in the Funariales and Bryales, but DNA sequence data has helped clarify the uniqueness of this moss, which is reflected in the establishment of the recently created order Pseudoditrichales, of which *P. mirabile* is the sole representative. *Pseudoditrichum* is a pioneer of moist open soils by late-lying snow beds in habitats where there is little competition from other plants due to

BELOW | The remarkable peristome of *Pseudoditrichum* showing the double ring of white peristome teeth.

DISTRIBUTION
Northwestern Canada and Arctic Siberia

ETYMOLOGY
Pseudo + *ditrichum*, referring to the superficial resemblance to the genus *Ditrichum*

NUMBER OF SPECIES
Only 1 known species

APPEARANCE
Plants small, $\frac{1}{16}$–$\frac{1}{8}$ in (1–3 mm) high. Leaves wider at the base then narrowing abruptly to a tapering point, costa single. Seta up to $\frac{1}{4}$ in (6 mm) long. Peristome double, with outer ring of teeth bright white and curved inward when dry. Vegetative reproduction with rhizoidal tubers and gemmae

HABITAT
On moist calcareous, silty soil

the harsh conditions so near the Arctic Circle. Its disjunct distribution suggests it is a relict species that has survived at these sites for an extremely long time and was once likely more widespread. Mosses at these high latitudes are particularly vulnerable to rapid global warming, which has the potential to dramatically alter their habitat conditions, and *Pseudoditrichum*, which represents such an isolated and distinct lineage of mosses, is likely to now be at risk.

ABOVE | Showing the habitat along the Eriechka River valley, Siberia, from where a single specimen of *Pseudoditrichum mirabile* was discovered in 2015.

RIGHT | The inconspicuous plants of *Pseudoditrichum* are distinguished from other similar mosses by their bright white peristome.

151

SCOULERIA

The *Scouleria* genus is a distinctive group of aquatic mosses that live in fast-flowing water. They are often submerged, but when water levels drop, they dry a characteristic jet-black color, clinging to exposed rocks. Most species of *Scouleria* have a thickened leaf margin, and this could be an adaptation to resisting damage to the plant in this turbulent habitat. The short, sturdy seta is also a likely adaptation to life in fast water. The genus is defined by some particular morphological features that, in combination, distinguishes them from other mosses. They lack a central strand in the stems; when present the peristome consists of 32 red teeth; there are some complexities to their capsule structure, with the capsule lid fused to a central column inside the capsule; and they produce relatively large and densely papillose spores.

There are six species accepted within *Scouleria*, though opinions have differed over how many distinct species can be defined within this genus. One of the taxonomic challenges when dealing with aquatic plants is that they often show a lot of morphological variation due to the high level of

LEFT | A beautiful rocky river ravine in Oregon, USA, where *Scouleria marginata* can be found growing on boulders.

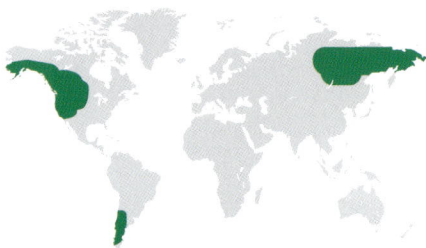

DISTRIBUTION
Northwestern North America and Northeast Asia, temperate South America and sub-Antarctic islands

ETYMOLOGY
After John Scouler (1804–1871), botanist and physician

NUMBER OF SPECIES
6 accepted species

APPEARANCE
Plants forming black or brown tufts, 1½–4 in (4–10 cm) tall. Leaves tongue-shaped to broadly lanceolate, costa single, margins smooth or toothed. Dioicous. Seta very short and stout. Capsule rounded becoming flattened with age, peristome present or absent. Calyptra hooded

HABITAT
Associated with rivers and streams, growing on soil banks or submerged rocks

ABOVE | The broad leaves of *Scouleria aquatica* are rounded at the apex and bright green at the shoot tips, darkening further down the stem.

BELOW | In this close-up of *Scouleria aquatica*, the capsule lid can be seen to be detached from the rim of the capsule mouth but it remains attached to a central column inside the capsule.

environmental factors affecting their growth (for example, changing water levels and the amount of light they are receiving). The relationships of *Scouleria* to other mosses has been ambiguous and for a long time *Scouleria* was classified in the Grimmiales. With more clarity now provided from DNA sequence data, the Scouleriales order is now recognized as distinct and includes the additional genera *Drummondia* and *Tridontium*.

BRYOXIPHIUM

BELOW | A close-up view of a turf of *Bryoxiphium madeirense*, showing its distinctly flattened shoots, on a rock wall at the edge of subtropical laurel forest in Madeira.

The genus *Bryoxiphium* could be mistaken for a common *Fissidens* moss at first glance as it shares the unusual habit of having its leaves distichously arranged up the stem and each leaf being folded lengthways. However, in *Bryoxiphium* the leaves are much more simple than in *Fissidens*. The stems tend to be unbranched, when growing they are a bright green and from a distance they can also resemble blades of grass.

B. norvegicum is a globally rare moss that has a particular strong hold in the Appalachian Plateau of eastern North America. This particular region remained ice-free during the last glaciation period. It is thought that *Bryoxiphium* was once more widely distributed but past glaciation events may have heavily influenced its current distribution, restricting it to the scattered sheltered spots where it was able to avoid the icy crush of the glaciers. Sporophytes are rarely produced, as the North American populations appear to be mostly female, so the plants' ability to disperse to a new location is very limited.

B. norvegicum grows on moist, vertical sandstone and can be abundant in the opening of caves with their own microclimates, which can vary substantially from the average conditions in the surrounding area. It is no surprise, therefore, that caves can often shelter rare disjunct moss species and relics that

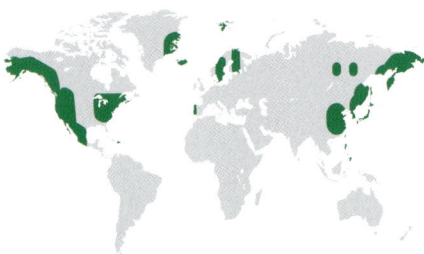

DISTRIBUTION
Widespread but disjunct; predominantly northern hemisphere

ETYMOLOGY
Greek *bryon* = "moss" + *xiphium* = "sword," referencing the growth form of the shoots

NUMBER OF SPECIES
4 accepted species

APPEARANCE
Plants green to brown, growing in tufts or pendant, ⅛–1¼ in (4–30 mm) long, shiny. Leaves oblong, lanceolate with blunt apex, folded lengthways, costa single. Dioicous. Seta straight or curved. Capsule often inclined, lacking a peristome. Calyptra hooded

ABOVE | This *Bryoxiphium norvegicum* is hanging down from a cave ceiling at a lake margin, Iceland.

RIGHT | This single leaf of *Bryoxiphium norvegicum* illustrates how the leaves are folded lengthwise in half.

have found a safe harbor among a wider changing environment. *B. madeirense* is an even rarer species found only on the tiny island of Madeira, where it is at risk of extinction as its populations are decreasing.

HABITAT
On moist soil banks and shaded cliffs

GRIMMIA

The *Grimmia* genus is the largest genus in the Grimmiales and is generally considered a "difficult" genus in terms of its taxonomy due to the large number of species and variability in their features. *Grimmia* is defined by its peristome of a single ring of divided teeth and some specific anatomical features of its leaf costa. Characters of the costa cells are very important in *Grimmia* taxonomy, and skills in cutting transverse sections of a leaf one cell thick soon become necessary to observe these.

Although the genus is widespread and found on every continent, almost half of the known species are endemics with narrow distribution. An accurate global picture of *Grimmia* species diversity still requires more taxonomic study and exploration

DISTRIBUTION
Widespread though not found in lowland tropical areas

ETYMOLOGY
After Johann Friedrich Karl Grimm (1737–1821), German physician and botanist

NUMBER OF SPECIES
108 accepted species

APPEARANCE
Forming cushions or mats, ¼–1½ in (5–40 mm) tall, green to black green. Leaves mostly lanceolate, often strongly concave or with a boat-like keel, costa often extends in silvery awn. Dioicous or monoicous. Seta short to long. Capsule usually ovoid with peristome of 16 teeth. Calyptra hood- or bell-shaped.

given the complexity of the group and the propensity for narrow endemics. New species of *Grimmia* are still steadily being described. For example, in 2023 two *Grimmia* species were described new to science from California, which is considered a particularly high-diversity area for bryophytes and especially *Grimmia*, with the suggestion that there may still be more species diversity in this region as yet undescribed.

   Grimmia species can thrive in a range of environments from temperate to extreme cold. They are typically associated with mountain terrain, where enthusiasts can be found hiking to remote ridges in search of rare species. However, some of the more widespread species are common in a variety of habitats, including urban settings, where instead of a mountain ridge they may perch on a stone wall top.

ABOVE | A close-up view of a cushion of *Grimmia orbicularis*, showing the curved setae keeping the immature capsules bent over into the leaves.

OPPOSITE | Neat, hoary cushions of *Grimmia montana* cling to rocks along with a community of colorful lichens in this scene from eastern Scotland.

RIGHT | The long white hairpoints typical of the genus can be seen in this cushion of *Grimmia sessitana*.

HABITAT
On rocks, usually acidic bedrock, usually in exposed habitats

RACOMITRIUM

The *Racomitrium* genus was traditionally characterized by its distinct leaf cells, which are elongate, thick-walled, and sinuose, and by the structure of the peristome, with linear teeth arising from a basal membrane. Taxonomic work, however, has revealed that this concept of *Racomitrium* does not reflect the evolutionary history of the group, and the genus was shown to actually be more heterogenous than realized. It is now split into four genera, with a narrower concept of *Racomitrium* plus *Codriophorus*, *Niphotrichum*, and *Bucklandiella*. The more tightly defined *Racomitrium* has long, white awns that extend from the leaf apex; the seta is roughened with minute papillae, a distinct feature for this genus; and the capsule bulges out to one side at the base.

R. lanuginosum is the most commonly encountered species in this genus. This large, distinctive moss is immediately recognizable with its long, white, toothed hairpoints, which are decurrent down the leaf margins. *R. lanuginosum* can form extensive thick mats in the Arctic tundra that dominate the landscape.

BELOW | The long, crinkly white hairpoints to the leaves are very distinctive in this dry shoot of *Racomitrium lanuginosum*.

DISTRIBUTION
Widespread in temperate and Arctic regions

ETYMOLOGY
Greek *rhakos* = "cloth rag" + *mítra* = "turban," referencing the calyptra frayed at the base

NUMBER OF SPECIES
5 accepted species

APPEARANCE
Large plants up to 6 in (15 cm) tall, brown/yellowish green, appearing gray from the white leaf awns. Leaves ovate to narrowly lanceolate, long white awn decurrent at leaf tip. Seta strongly papillose. Capsule ovoid, with peristome teeth long and split nearly to the base. Calyptra bell-shaped or hooded

HABITAT
On soil or rock in dry, exposed areas

CAMPYLOPUS

This large genus of robust mosses often grow as pioneers of open or disturbed habitats. The anatomy of the leaf costa is a critical feature in *Campylopus* taxonomy. The position of the thick-walled bundles of stereid cells that provide structure support in relation to other cell types in the costa can help identify a species.

C. introflexus is an invasive species native to the southern hemisphere that has been introduced to Europe and western North America. Following its arrival in new continents *C. introflexus* has spread rapidly. In the Netherlands it is known as *tanknos* due to the likely possibility that tanks during World War II played a role in spreading populations. Its success is linked to its ability to colonize disturbed areas of ground very quickly and disperse by fragmented leaves as well as by spores. The extensive mats that can be formed by *C. introflexus* can have a negative impact in areas where they encroach on more delicate habitats, such as coastal dunes in the Netherlands.

ABOVE | Species of *Campylopus* often have the setae arched over like a swan's neck, as shown in this photo of *Campylopus flexuosus*.

DISTRIBUTION
Widespread

ETYMOLOGY
Greek *campylos* = "curved" + *pous* = "foot," referring to the curved seta

NUMBER OF SPECIES
185 accepted species

APPEARANCE
Medium to large plants 1¼–4 in (3–10 cm) tall. Leaves narrowly lanceolate, costa filling more than half of leaf width. Dioicous. Seta curved downward when immature. Capsule often small, swelling at side of base, peristome teeth each divided in 2. Calyptra hood-shaped

HABITAT
Usually on soils or gravel, occasionally on rock, always on acidic substrates

LEUCOBRYUM

BELOW | Shoots of *Leucobryum acutifolium* from South Africa, showing the thick whitish-green leaves typical of this genus.

P ale, succulent cushions of *Leucobryum* grow in dense, rounded mounds over the woodland floor. It can occasionally form extensive pure carpets to great effect, and it is no wonder that *Leucobryum* is often a feature in Japanese moss gardens. While a common feature of temperate forests, most of the species' diversity is found in tropical regions.

The leaf structure in *Leucobryum* is formed from two kinds of cells. There are two or more layers of large, empty, colorless cells with pores in the cell walls called leucocysts, whose function is to absorb and hold a large amount of water. Sandwiched between the leucocysts is a central layer of small, green cells called the chlorocysts. Another interesting biological feature of *Leucobryum* is its reproductive strategy. The species are dioicous but the male plants are so tiny that they grow on the leaves of the female plants. This useful adaption ensures proximity of male and female plants for successful fertilization and subsequent sporophyte production.

Leucobryum is such a pleasingly attractive and tactile moss that it has become popular among

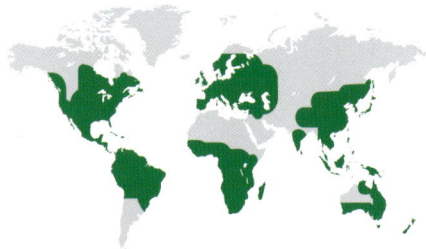

DISTRIBUTION
Widespread in temperate and tropical regions

ETYMOLOGY
Greek *leukós* = "white" + *bryon* = "moss," alluding to the pale color of the plants

NUMBER OF SPECIES
80 accepted species

APPEARANCE
Plants white to pale green, in cushions, ¼–6 in (0.5–15 cm) tall. Leaves with narrow sheathing base and lanceolate to ligulate spreading limb, an expanded costa comprising leucocyst and chlorocyst layers. Dioicous with dwarf males. Seta straight. Capsule with 16 peristome teeth. Calyptra hooded

HABITAT
On dry soils and occasionally on rocks, commonly in woodlands

ABOVE | *Leucobryum glaucum* forming wonderfully tactile, dense, rounded hummocks on a woodland floor.

terrarium enthusiasts. In fact, once you get the feel for recognizing *Leucobryum*, these mosses can often be spotted in unexpected places such as artificial planting schemes and naturalistic decorations. They cannot survive in conditions where they may dry out, so dead cushions are often dyed green before being used in displays. The hand-sized cushions can take many years to grow, and wild harvesting can seriously threaten these natural populations, which are best appreciated when happened upon in a beautiful woodland glen.

PLEUROPHASCUM

The genus *Pleurophascum* is a bryological curiosity so unique that when first described in the scientific literature in 1875 it was hailed as "of no less interest to the museologist than is *Rafflesia* or *Welwitschia* to the phanerogamist"! Of particular note are the sporophytes, which develop on the sides of the stem, and the remarkable large, spherical capsules that have no lid and are often brightly colored. The broad, rounded leaves have no costa and are concave.

There are only three known species of *Pleurophascum* and all have very narrow distributions. The aptly named *P. grandiglobum* is endemic to Tasmania. It has large, round capsules that are indeed grand globes at ⅛–¼ in (3–6 mm) in diameter. *P. ovalifolium* from New Zealand has capsules that become disc-shaped and a bright orange-red color when mature. *P. occidentale* of western Australia produces its sporophytes at the tips of branches coming off the main stem rather than laterally off the side of the stem, and the large capsules only just sit above the leaves.

LEFT | A clump of *Pleurophascum grandiglobum* in Tasmania, the bright-yellow, globe-shaped capsules standing out among the darker vegetation.

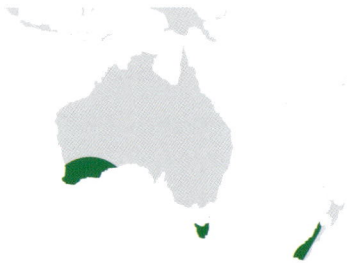

DISTRIBUTION
Australia and New Zealand

ETYMOLOGY
Greek *pleuron* = "side" or "rib" + physce = "bladder," alluding to the capsules of *P. grandiglobum* arising on short side branches

NUMBER OF SPECIES
3 species

APPEARANCE
Main stems creeping underground with tiny appressed leaves, secondary stems upright to 1¼ in (3 cm) tall and leafy. Leaves concave, crowded, appressed against stem, costa absent. Dioicous. Seta short or elongate. Capsule large, green, yellow, or orange, round or disc-shaped, lacking both lid and peristome

Given the lack of a capsule lid, the method of spore dispersal has been perplexing in this genus. Capsules of this kind usually rely on the capsule wall disintegrating to allow spores to be released. The capsules of *Pleurophascum* in Tasmania have been routinely observed nibbled by unknown creatures, and it is possible that invertebrates play a role in spore dispersal. Another theory is that the whole capsule could be a dispersal unit floating on water in the boggy habitats favored by *Pleurophascum* in New Zealand, where they could feasibly travel in the water and release spores at new sites.

ABOVE RIGHT | A close-up of *Pleurophascum grandiglobum* capsules with their tiny calyptrae.

RIGHT | The pale-orange capsules of the Western Australia endemic *Pleurophascum occidentale* also demonstrate the lack of a lid.

HABITAT
On soil in heaths or forests or by boggy pools

EUSTICHIA

U ntil recently the Eustichiaceae family was classified in the Dicranales, but along with several families in the Dicranales it has recently been elevated to the status of order to reflect its distinct evolutionary lineage. *Eustichia* is characterized by its stems, which are often very long with the lower part buried in the soil. The upper region of the plant is rather flattened and the leaves are folded lengthways, making *Eustichia* appear distichous, and the leaf costa extends beyond the tip of the leaf with a slight curve, giving the leaf tip a hook-like appearance.

There has been some uncertainty over the number of distinct species represented by this genus, and until recently the consensus was that there was just one variable species: *E. longirostris*. In 2022 researchers re-examined critical scientific specimens of *Eustichia*, in particular the nomenclatural types housed in herbaria that represent the physical specimens bound to a species name. The *Eustichia* plants occurring in southern Africa had previously been known under the name *E. africana*, but since 1923 bryologists considered these two names to

LEFT | Looking at a leaf of *Eustichia longirostris* under the microscope clearly illustrates how the small, oval leaves are folded lengthwise.

DISTRIBUTION
Mexico to South America, also southern Africa and the Southern Ocean islands

ETYMOLOGY
Greek *eû* = "well, good" + *stíkhos* = "line, row," alluding to the distichous appearance of the plants

NUMBER OF SPECIES
2 species

APPEARANCE
Forming light-green tufts up to ¼ in (5 cm) tall, flattened, and appearing distichous. Leaves dimorphic, scale-like in lower shoot, 2 ranked above with leaves folded lengthways, leaf cells papillose. Dioicous. Seta elongate. Capsule usually ribbed, peristome single ring of 16 teeth. Calyptra hood-shaped

represent plants belonging to the same species, and the name *E. africana* was lost to synonymy. This new research suggests *E. africana* may be a distinct species after all, and the names have changed again. Taxonomic research like this is important because it describes species diversity in as accurate a way as possible based on current information. The conservation implications for the African plants changes significantly if they are a rare and endemic species rather than one that is more widespread.

HABITAT
On soil or soil-covered rocks

BELOW | *Eustichia longirostris* has a widespread distribution across Central and South America.

ABOVE | The flattened shoots and folded leaves with curved tips are distinctive of *Eustichia longirostris*.

AMPHIDIUM

The *Amphidium* genus typically forms conspicuously large, dense cushions on shaded, damp rock faces in montane areas. The long, narrow leaves are tightly twisted when dry, then spread out when wet. The capsules of *Amphidium* are characteristically cup-shaped with a wide mouth lacking a peristome and ribbed sides to the capsule, though unhelpfully *Amphidium* plants are often sterile. Deciphering the evolutionary relationships of *Amphidium* has been challenging, largely because of the absence of a peristome, the features of which were traditionally used to classify mosses into related groups. The leaf anatomy also does not offer many distinctive features to clearly separate them from other small acrocarpous mosses. It was most recently placed within the Dicranales but has now been established within its own order, the Amphidiales.

There are 11 species recognized at the moment but it is possible that further scrutiny will reveal several of these to be superfluous, since some are based on sterile specimens that may actually belong to different genera. Another taxonomic problem is

LEFT | *Amphidium lapponicum* is a widespread species, photographed here in Iceland, which grows in damp crevices of cliffs.

DISTRIBUTION
Widespread though predominantly northern hemisphere

ETYMOLOGY
Latin *amphora* = "large wine vessel," referring to the shape of the capsule

NUMBER OF SPECIES
11 accepted species

APPEARANCE
In dense lime-green to dark-green turfs or cushions up to 2 in (5 cm) tall. Leaves lanceolate to lingulate, margins recurved below, serrate above, costa single, crisped when dry, leaf cells papillose. Seta short. Capsule pear-shaped, flared at opening. Peristome absent

HABITAT
On soil or rocks

that some of the critical historic type specimens upon which names are based have not been located, and the original descriptions are not detailed enough to confirm their identity. Based on molecular DNA sequence data, the diversity of *Amphidium* is suspected to be around six species, but further taxonomic work is needed to resolve all the names listed under this genus. The highest diversity of *Amphidium* is found in western North America, Macaronesia, and Central Asia.

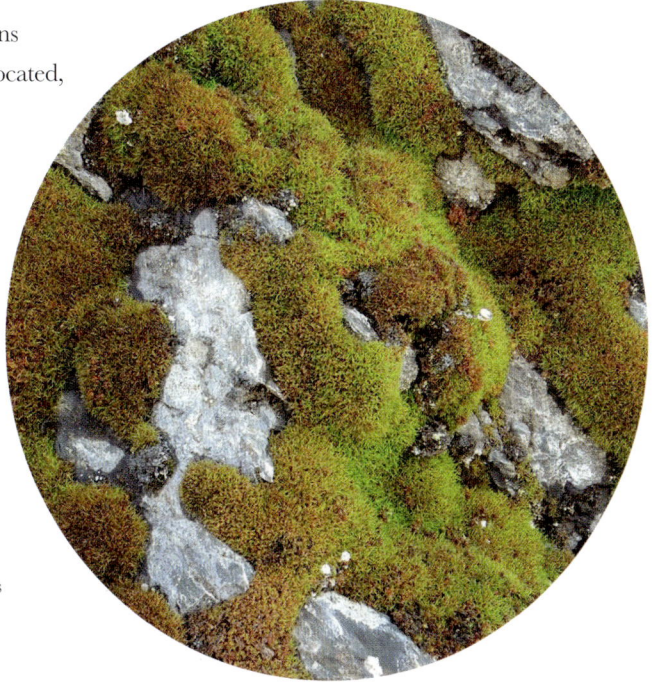

BELOW | The individual shoots of this *Amphidium mougeotii* are not particularly distinctive but *Amphidium* can be immediately recognizable when forming its dense lime-green cushions on damp rock faces.

RIGHT | *Amphidium* typically forms dense cushions over damp rocks, as shown by this *A. californicum*, which is endemic to coastal western North America.

DICRANUM

The Dicranales represents a diverse group of mosses, and with 124 genera classified within the order, it is second only in size to the Hypnales. It has traditionally been very broadly defined, grouping families that lacked the more clearly defined traits of the closely related Grimmiales and Pottiales. More recent research, supported by DNA sequence data, has separated out some families into their own orders and left a more narrowly defined Dicranales.

Dicranum species are recognized by their long, narrowly tapered leaves that often sweep majestically to one side. Particular features that separate *Dicranum* from other similar-looking members of the Dicranales include the leaf cells, which are quadrate throughout and rather enlarged in the basal corners as alar cells. The costa is single and quite narrow compared to the broad costa that fills the leaf in some other genera. Many *Dicranum* species produce dwarf male plants that are so tiny they perch on the leaves of female plants, solving the problem of proximity that dioicous species often face.

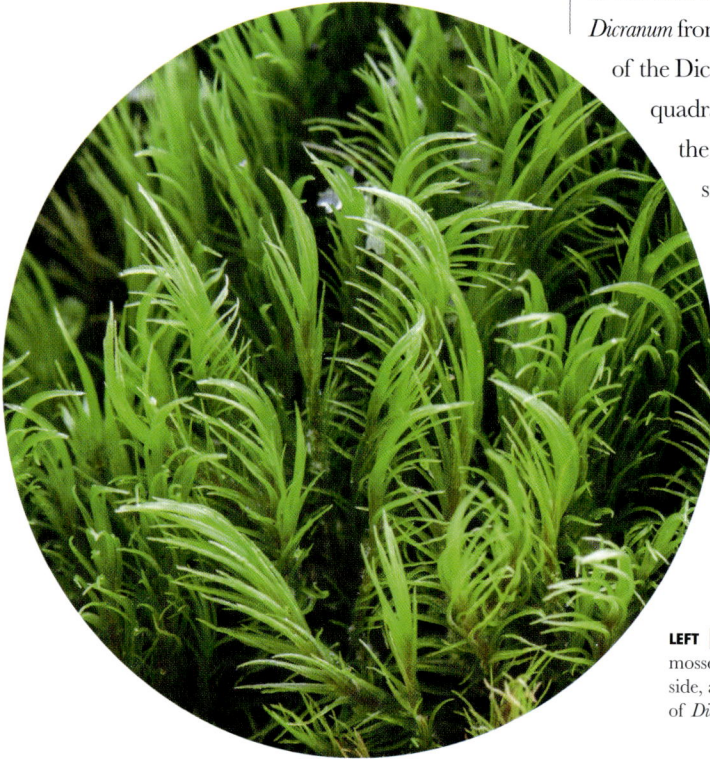

LEFT | The leaves of *Dicranum* mosses are often swept to one side, as show by these plants of *Dicranum montanum*.

DISTRIBUTION
Widespread

ETYMOLOGY
Greek *dikranos* = "pitchfork," referring to the divided peristome teeth

NUMBER OF SPECIES
94 accepted species

APPEARANCE
Plants forming medium to large dense tufts, usually ¾–4 ¾ in (2–12 cm) long. Leaves straight to curved, costa single and quite narrow. Dioicous. Capsule erect or incline, peristome a single ring of 16 teeth, usually divided ⅓–⅔ of their length

HABITAT
Mostly on soil on woodland floor but also on rotting logs, epiphytes on trees and occasionally rocks

CALYMPERES

Mosses of the *Calymperes* genus are typically found in lowland tropical forests. Their calyptra is unique, staying permanently attached to the capsule. The top of the calyptra tightens around the capsule lid beneath it and in dry conditions lifts the lid away from the capsule mouth, allowing spores to be released through tears that open up in the sides of the calyptra. When wet, the calyptra twists and closes the spores' escape routes. *Calymperes* also often produce long cigar-shaped gemmae on specialized leaves, clustered together looking like spiky green balls at the leaf tips.

Under the microscope *Calymperes* can be identified by the patterns of cells in its leaves. Most species have a unique intramarginal band of clear, linear cells called a "teniola," which can be seen forming a border along the basal edge of the leaf but which is positioned within a short space from the actual leaf edge. The presence of this band of cells distinguishes *Calymperes* from the closely related *Syrrhopodon*, with which it can sometimes be confused.

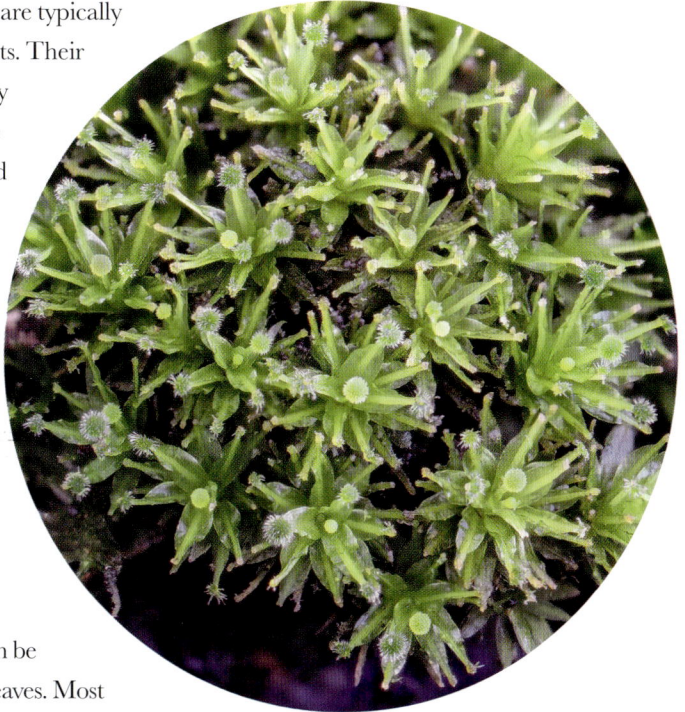

ABOVE | The tropical *Calymperes tenerum*, photographed here on Réunion Island, produces spiky gemmae balls on the ends of its leaves.

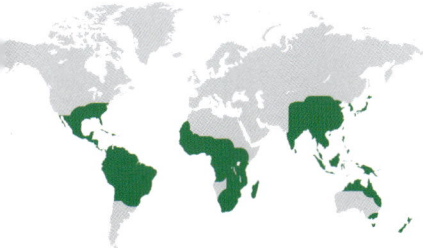

DISTRIBUTION
Widespread across tropical regions

ETYMOLOGY
Greek *kalymma* = "veil" + *peiro* = "to pierce," alluding to the calyptra covering the capsule, with tears through which spores are released

NUMBER OF SPECIES
47 accepted species

APPEARANCE
Plants up to ⅜ in (1 cm) tall. Leaves often dimorphic, with some modified to produce gemmae, small rounded green cells in upper leaf and large clear cells below, occasionally with intramarginal border. Dioicous. Capsule with peristome absent. Calyptra persistent, covering capsule, with vertical slits when dry

HABITAT
Usually epiphytic

FISSIDENS

The leaves of *Fissidens* mosses have a complex structure that is unique in the moss world. Each leaf has a pocket formed of two sheets of leaf tissue at the base that creates a slot that the leaf above it on the stem fits into. To fully describe this leaf structure, it is useful to imagine how a model of such a leaf could be constructed. Starting with a basic leaf shape with a costa running up the middle, the leaf would then be folded lengthways along the line of the costa. If the upper part were to fuse together as a flat blade, this would leave an open "pocket" in the lower half. If then an additional wing of leaf tissue was added to the back of the costa, this would result in a typical *Fissidens* leaf.

In the late 1790s *Fissidens* was credited with saving the life of explorer Mungo Park, who journeyed across western Africa in search of the Niger River. At a particular low point, Park was robbed and left naked and alone in the bush. He then noticed a beautiful small moss and reflected that if a God had taken such care to create this exquisite tiny plant of seemingly little importance, then surely he would also be supported in this time of crisis. With this realization, Park was able to continue the journey to the safety of the coast. This "motivational moss" was later described as a species new to science, *F. parkii*.

BELOW | This single leaf of *Fissidens osmundoides* illustrates the unique structure of the leaf, with the "pocket" of folded leaf blades in the lower half.

DISTRIBUTION
Worldwide

ETYMOLOGY
Latin *fissus* = "split" + *dens* = "tooth," referring to the split peristome teeth

NUMBER OF SPECIES
478 accepted species

APPEARANCE
Plants tiny to rather large, shoots up to 4 in (10 cm) long. Leaves distichous, lower leaf blade a boat-shaped sheath with rest of blade undivided, costa single. Mostly monoicous. Seta short or elongate. Capsule erect or inclined, peristome a single ring of 16 teeth, usually divided 1/3 – 2/3 of their length

ABOVE | The African *Fissidens ovatus* with flattened shoots and long folded leaves typical of the genus.

RIGHT | The European species *Fissidens gymnandrus*, showing the leaves in two rows up the stem and the folded leaf structure.

HABITAT
Shaded soil, occasionally rock, can occupy aquatic habitats

SCHISTOSTEGA

Most species of mosses around the world do not have common names, but occasionally a moss comes along so charismatic and intriguing that several names are acquired—Luminous Moss, Goblin's Gold, and Dragon's Gold all refer to *Schistostega pennata*. These names are woven around *Schistostega's* most fascinating feature: its ability to seemingly glow in the dark.

Schistostega is not able to compete well with other plants for space, so it has adapted to grow in deeply shaded places where its competitors cannot follow. The first stage in a moss life cycle after spore gemination is to form a mat of filaments, the protonema, from which leafy gametophyte shoots develop. In *Schistostega* this protonema can grow extensively, forming large mats. These filaments contain chloroplasts to undertake photosynthesis, and they have a clever trick to make the most of the limited light available to them. The filaments have specialized cells that contain a large lens-shaped structure whose curved surface is able to focus light onto its chloroplasts. Light not absorbed by the chloroplasts is bounced back out of the cell, creating the luminous effect of sparkling green glitter. Growing in dark crevices, wind power cannot be relied upon for dispersal, so *Schistostega's* spores have adapted to be sticky and dispersed by animals.

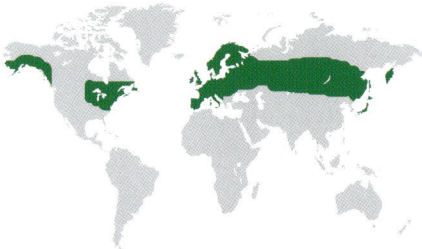

DISTRIBUTION
Northern hemisphere of North America, Europe, Russia, China, and Japan

ETYMOLOGY
Greek *schistos* = "divided" + *stego* = "cover," alluding to an incorrect early observation that the operculum of the capsule splits

NUMBER OF SPECIES
Only 1 species

APPEARANCE
Persistent protonema reflective, producing a green-gold color. Leafy plants gray-green, small, up to $\frac{1}{16}$ in (1.5 mm) long, shoots distichous and flattened. Dioicous. Seta short. Capsule ovoid, peristome absent. Calyptra bell-shaped

OPPOSITE | *Schistostega* has an unusual flattened leaf structure with the leaves lacking a costa and fused at the bases.

ABOVE | The leafy shoots of *Schistostega pennata* and its luminous protonemata, photographed in Cornwall, England.

Schistostega is widespread across the northern hemisphere, though it is becoming a rare moss across much of its range. It is threatened by both human disturbance of its cave habitats and the combination of higher temperatures and lower humidity experienced by some regions due to global warming.

HABITAT
On deeply shaded soil such as around cave entrances and animal burrows, or under overhanging soil banks

GLYPHOMITRIUM

The family Rhabdoweisiaceae was classified within the Dicranales up until 2023, when it was assigned to a new order along with the Rhachitheciaceae family. The Rhabdoweisiales are a diverse group of mosses that are united by molecular DNA data but have few defining morphological characters. They tend to be small,

BELOW | *Glyphomitrium*, such as this *G. daviesii*, is recognized by its delightful capsules with paired peristome teeth spreading out around the capsule mouth.

DISTRIBUTION
Eastern Asia and Europe

ETYMOLOGY
Greek *glyphē* = "carved work" + *mítra* = "turban," referencing the calyptra, which is longitudinally pleated

NUMBER OF SPECIES
12 accepted species

APPEARANCE
Plants forming tufts ¼– ⅝ in (0.5–1.5 cm) tall. Leaves twisted when dry, spreading when wet, lanceolate, costa single. Monoicous. Capsules short-cylindrical or cup-shaped, peristome a bright-orange or red ring of 16 teeth that are often paired. Calyptra covering capsule, pleated, lobed at base

HABITAT
Epiphytic or on rocks in humid sites

cushion-forming acrocarpous mosses that usually grow on trees in cool, damp climates.

The capsules of *Glyphomitrium* are particularly pretty, with a red-orange peristome of 16 teeth paired together to look like 8 large teeth, which are reflexed when dry. This distinctive peristome, along with the bell-shaped calyptra covering the capsule, help define the genus. *Glyphomitrium* is also known for its large spores, which are endosporic, meaning the spore cells start to divide while still in the capsule.

Glyphomitrium is a genus of mainly Asian distribution, with nine species known from its region of peak diversity in China. One species, *G. daviesii*, occurs in Europe, where it requires an oceanic climate of cool summers and mild winters, usually favoring a coastal location. *Glyphomitrium* was for a short time recorded from North America, but after some deliberation bryologists decided that the historic specimens that these records were based on had been mislabeled and were probably European in origin. The East Asian species tend to be epiphytic, while the European *G. daviesii* is restricted to rocks and is thought to represent an early diverging lineage within the genus.

ABOVE | This European *Glyphomitrium daviesii* tends to grow on rocks while the Asian species grow mainly as epiphytes on trees.

SORAPILLA

The genus *Sorapilla* is one of the rarest and most unusual mosses in the world. Its boat-shaped leaves are flattened and arranged in opposite rows up the stem. One side of the leaf forms a pocket of similar structure to the *Fissidens* mosses, which is extraordinary since these two genera are not closely related. The leaf cells along the margin of this pocket form a border of smooth, clear, and elongate cells, while the adjacent leaf cells are smaller, with papillose cell walls. The sporophytes develop laterally from the sides of branches, not terminally from the ends of the shoots, grouping *Sorapilla* with the creeping pleurocarpous mosses.

This elusive moss was first discovered in 1857 by botanist Richard Spruce during his expedition to the Amazon. Spruce's collections of *S. sprucei* remain the only record of this species, since it has never been re-found. In 1900 a second species was described, *S. papuana*, based on specimens from Papua New Guinea and later found in Queensland, Australia, and in Indonesia. *Sorapilla* is so rare that our knowledge of this plant is based on a handful of largely historic herbarium specimens.

In 2015 a new population of *S. papuana* was found by a botany student on a field trip to a remote rainforest in northern Queensland, and this

DISTRIBUTION
Ecuador, Indonesia, northwestern Australia, and Papua New Guinea

ETYMOLOGY
The name *Sorapilla* is based on the word for "moss" as spoken by the people of the eastern Andes, whom Spruce encountered on his expedition

NUMBER OF SPECIES
2 accepted species

APPEARANCE
Shoots whitish-green, irregularly branched, up to 1 ¾ in (4.5 cm) long. Leaves with doubled lamina on one side creating a pocket. Dioicous. Seta short. Capsules immersed, peristome double with an outer ring of teeth and an inner ring forming a low wall

exciting discovery enabled detailed analysis of this remarkable plant based on fresh material. DNA sequence data has now confirmed the taxonomic placement of *Sorapilla* as an isolated group within the broader Dicranidae class, and in 2023 it was elevated to its own order.

HABITAT
Epiphytic on trees

ERPODIUM

The Erpodiales are small pleurocarpous mosses found in tropical and subtropical zones. They tend to occur in arid or seasonally dry regions. Taxonomic revision has reallocated many species that were classified in *Erpodium* into other genera, such as *Venturiella*. *Erpodium* is characterized by the papillae on the leaf cell wall that are both O- and C-shaped when viewed under the microscope; the capsules, which only just emerge above the perichaetial leaves (that are also notable for not being much larger than the vegetative leaves); and the absent peristome and the bell-shaped calyptra. *Erpodium* now includes just two rather similar species. *E. cubense* is known only from Cuba and the Dominican Republic. *E. domingense* is more widespread across Central America and differs by its smaller leaf cells and capsules that are held a little higher above the leaves. These small, creeping mosses with their oval leaves are often mistaken for leafy liverworts in the field.

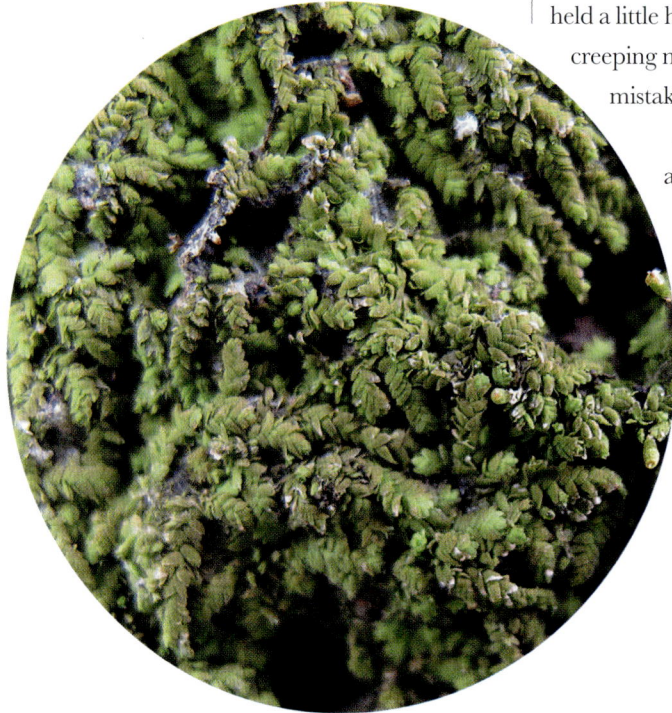

Erpodium has a very narrow distribution, and in 2012 it was observed for the first time in suburban Miami, Florida, on ornamental palms. The Florida moss flora is well known, and these mosses were only found in one location, which suggests this species was introduced, raising the question of how it arrived. The Date Palms

LEFT | The small creeping shoots of *Erpodium* with their oval leaves have a superficial resemblance to leafy liverworts at first glance.

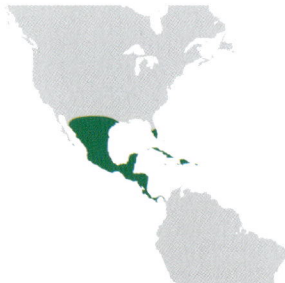

DISTRIBUTION
Restricted to the central region of the American continent

ETYMOLOGY
Greek *erpo* = "creeping," alluding to the growth habit of the plant

NUMBER OF SPECIES
2 accepted species

APPEARANCE
Plants very small, creeping, and irregularly branched, up to ⅜ in (1 cm) long. Leaves mostly ovate, concave, with costa absent, leaf cells papillose. Monoicous. Seta short. Capsule short-cylindrical, peristome absent. Calyptra smooth or plicate

HABITAT
Epiphytic on trees, occasionally on logs or rocks, in dry forests

planted into this landscaped area originated from California and Arizona, where *Erpodium* does not occur, so this ruled out the possibility of the mosses inadvertently hitching a ride with the transplanted palms. Scientists concluded that these populations had instead probably been introduced to Miami from the Caribbean by tropical storms.

BELOW | Plants of *Erpodium domingense* growing on the trunks of Date Palm, *Phoenix dactylifera*, on the Florida International University campus, Miami, Florida.

TREMATODON

The Bruchiales is a newly established order that includes just the Bruchiaceae family of five genera. It is morphologically characterized as a group of acrocarpous mosses that have capsules with a well-developed neck and peristome that, when present, is formed of a single ring of teeth.

Trematodon has short, scale like-leaves at the base of the stems with larger leaves further up the stem that have a sheathing base and narrowly taper from a wide base. This leafy gametophyte does not vary a great deal between different *Trematodon* species, and mature sporophytes are usually needed for an accurate identification. *Trematodon* is closely related to *Bruchia* and the two genera could be confused. *Trematodon* can be distinguished by its longer setae, which raise the capsules up above the leaves, and its capsules, which have a lid to release spores—as opposed to *Bruchia*, whose capsules have no lid, with spores being dispersed through a disintegrating cell wall.

Trematodon is in need of taxonomic scrutiny. A large number of species have been described in this genus over the last 200 years, but these will not all relate to distinct species and a degree of synonymy within these names is expected. A diversity of around 25 species has been suggested as a likely outcome of further taxonomic research. One species, *T. perssoniorum*, is so rare that it is known from a single locality in the Azores. This species is considered at critical risk of extinction due to its tiny distribution and the declining quality of its wetland habitat.

LEFT | *Trematodon* plants can be recognized by the long, expanded capsule neck, as shown by this *T. ambiguus*.

DISTRIBUTION
Worldwide

ETYMOLOGY
Greek *tremato* = "perforated" + *odon* = "tooth," referring to the perforated peristome teeth

NUMBER OF SPECIES
76 accepted species

APPEARANCE
Small tufts up to $\frac{1}{8}$ in (4 mm) high. Leaves tapering abruptly to a narrow point from a wide base, leaf cells quadrate or short-rectangular. Monoicous. Seta elongate. Capsule cylindric with a strongly differentiated neck, peristome absent or present with 16 simple, perforated or forked teeth. Calyptra hood-shaped

ABOVE | The beautiful *Trematodon pillansii*, a species endemic to South Africa.

RIGHT | *Trematodon* species are colonizers of bare soil, as shown here by the widespread *T. longicollis*, photographed in Taiwan.

HABITAT
Growing on exposed soil, often in disturbed sites

CERATODON

The Ditrichales includes 22 genera of small acrocarpous mosses with mostly narrow tapering leaves with a costa and capsules that usually have a single ring of peristome teeth. *Ceratodon purpureus* is the most widespread species in the genus and occurs on every continent. It is a species notorious among bryologists for looking like something else when spotted out in the field—potentially something more interesting! Many a specimen has been collected optimistically only for the true identity of *C. purpureus* to be revealed under the microscope.

One of the challenges in taxonomy comes from establishing whether variability among specimens is the result of variable environmental conditions that cause the plants to express a different morphology or from genetic differences that correspond to a distinct evolutionary unit that we would identify as a species. In 2018 scientists published a study investigating *Ceratodon* diversity in southern Europe to test whether the known variability of this genus could be disguising genetic distinctions. They undertook both genetic analysis and cultivation experiments. By growing plants in the laboratory, they confirmed that *Ceratodon* did alter its expression of certain morphological features under differing conditions. The study also identified plants in southern Spain as

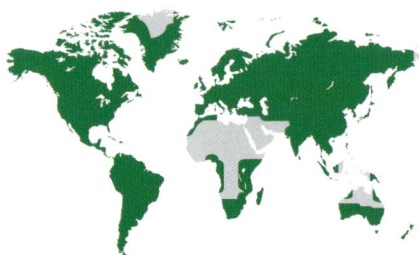

DISTRIBUTION
Widespread

ETYMOLOGY
Greek *keratos* = "horns" + *odon* = "teeth," alluding to the forked peristome teeth

NUMBER OF SPECIES
5 accepted species

APPEARANCE
Plants forming tufts or mats, green but often tinged reddish brown, up to 1¼ in (3 cm) tall. Leaves contorted when dry, narrowly triangular, margins recurved, costa single, leaf cells smooth, quadrate or short-rectangular. Dioicous. Seta elongate, red, purple, or yellow. Capsule with peristome a ring of 16 teeth, each split nearly to base. Calyptra hooded

ABOVE | The mature sporophytes of *Ceratodon purpureus* produce a dramatic show of reddish-purple setae.

HABITAT
On open soil or rock, often in disturbed sites

being both genetically and morphologically distinct from any other *Ceratodon*, resulting in the description of the new species *C. amazonum*. In a further twist to the tale, the previously recognized species *C. conicus* was found to be a hybrid of these two species. Perhaps we should all be looking more closely at *Ceratodon* after all.

BARBULA

T he Pottiales is a diverse order of cushion-forming mosses known for their desiccation tolerance and adaptation to living in harsh environments. It has long been a well-supported natural grouping united by morphological and molecular DNA data. There are 108 genera classified within this order and well over 1,000 species, with the highest diversity in the Mediterranean. These are mosses that may go unnoticed in periods of drought, when their leaves curl and twist up and they can take on a rather dull appearance. As soon as rain arrives, however, their leaves spread out and the plump green cushions swell with water.

Barbula plants are small, neat-looking, and often a characteristic yellow-green. *Barbula* gets its name from its long peristome formed of 32 filamentous teeth that are twisted when dry and joined in a basal membrane below. When *Barbula* was first officially described over 200 years ago, it had a much broader definition, encompassing many mosses that have since been moved into other genera based on new taxonomic evidence. For example, many *Barbula* species have since been reclassified

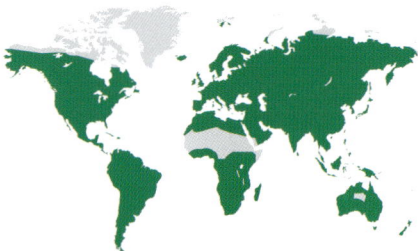

BELOW | The densely papillose leaf cells of *Barbula unguiculata*.

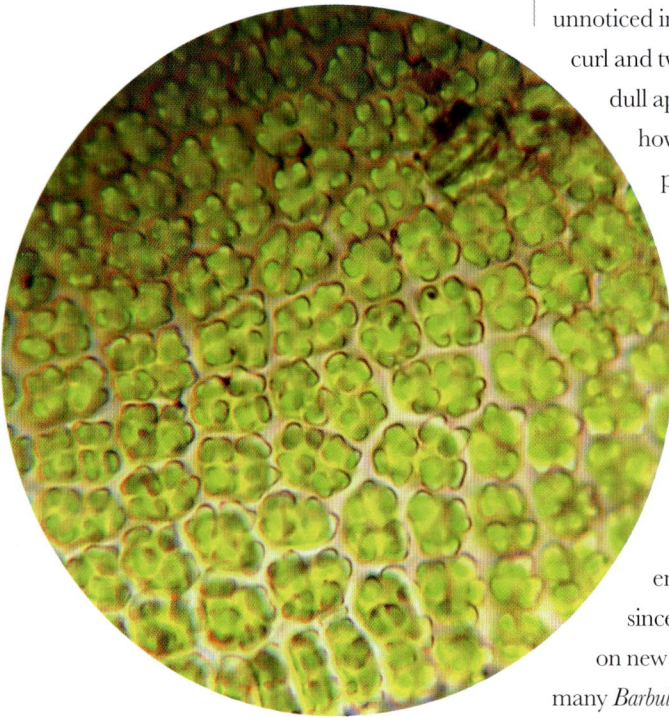

DISTRIBUTION
Widespread, mostly in temperate regions

ETYMOLOGY
Latin *barba* = "beard" + *ula* = "small one," referring to the long, twisted peristome

NUMBER OF SPECIES
101 accepted species

APPEARANCE
Plants often yellowish green, forming cushions up to ¾–1½ in (2–3 cm) high. Leaves contorted when dry, spreading when moist, broad or narrowly tapering, costa single and usually extending beyond leaf tip, upper leaf cells papillose. Dioicous. Seta short to elongate. Capsules with 32 filamentous teeth, strongly twisted. Calyptra hooded

HABITAT
On soil and rocks, rarely epiphytic

within *Didymodon*, a genus with which current *Barbula* plants could still be confused. Even more species were removed from *Barbula* in a study in 2013 that used molecular DNA data to show that the concept of *Barbula* at this time was not monophyletic, meaning the group did not share one common ancestor. Several new genera were duly established to create a taxonomy that is more reflective of their evolutionary past.

BELOW | *Barbula unguiculata* plants form distinctly yellowish-green tufts, often on bare soil as shown here.

RIGHT | The extravagant peristome of *Barbula* has led to a common genus name of the "Beard Mosses."

SYNTRICHIA

Cushions of *Syntrichia* are usually recognized by their broad upper leaves that narrow to a slim base and typically have a long, white hairpoint extending from the tip. Microscopically, each leaf cell wall may have several papillae, and these may be branched several times. This minute sculpturing is best appreciated under a powerful scanning

BELOW | Dry cushions of *Syntrichia caninervis* on a parched, sun-baked rock in Colorado, USA.

DISTRIBUTION
Widespread, though mainly temperate

ETYMOLOGY
Greek *syn* = "plus" + *trichos* = "hair," alluding to the twisted peristome

NUMBER OF SPECIES
100 accepted species

APPEARANCE
Plants often yellow/reddish green, forming tufts, ¼–¾ in (5–20 mm) high. Leaves twisted when dry, spreading when moist, usually ovate, ligulate, or spatulate, and ending in hairpoint, upper leaf cells papillose, lower leaf cells larger and clear. Dioicous or monoicous. Seta elongate. Capsules cylindric, peristome 32 spiraled teeth. Calyptra hooded

electron microscope but can also be effectively viewed by cutting a thin section of the leaf and viewing it side-on under a standard light microscope. *Syntrichia* is a great example illustrating the adaptive structures that enable water to be manipulated over the leaf surface as it is channeled through this network of papillae via capillary action.

In 2024 *Syntrichia* made news headlines when scientists in China claimed to have found a moss that could survive on Mars. *S. caninervis* is found naturally in such challenging terrain as Antarctica or the Mojave Desert in California. It can survive drought, extreme cold, and high levels of radiation. This study tested the limits at which this moss can survive as part of research into the potential for future planetary colonization. *Syntrichia* may not seem the most useful plant to take to Mars given that it is inedible—but it is possible that mosses could deliver other useful services in space as they do on Earth, stabilizing rocky terrain and enabling other plants to grow. While the reality of terraforming Mars is still a distant dream, this study has certainly highlighted the amazing biological potential of this tiny desert moss, crowning *Syntrichia* as Earth's toughest plant.

LEFT | Some species of *Syntrichia*, like this *S. montana*, have long silvery hairpoints and a leaf with a slightly constricted waist.

HABITAT
On soil or rocks or epiphytic, often in dry, exposed habitats

HEDWIGIA

The Hedwigiales order comprises the moss families Hedwigiaceae, Helicophyllaceae, and Rhacocarpaceae. The Hedwigiaceae includes four rather similar genera. These are fairly robust mosses with a creeping habitat and sporophytes that develop at the tips of the branches. They are technically acrocarpous mosses due to the terminal position of the sporophytes on the short lateral branches, though the branching pattern is not typical of this group. This growth form can also be found in the Orthotrichales and Grimmiales orders.

Typical species of *Hedwigia* are quickly recognized by the white-tipped leaves of the dark-colored plants. In all but one species of *Hedwigia* the perichaetial leaves that protect the developing sporophytes have a ciliate margin, which is very unusual in mosses. The short seta keeps the smooth capsules immersed within the leaves. The leaves are lacking a costa and the cell walls of the leaf lamina are thick walled with a wavy margin and ornamented with multiple small papillae.

For nearly 200 years the genus *Hedwigia* included just one easily recognized species. In the 1990s a more critical review of *Hedwigia* revealed that more than one species was present in Europe, and in the following years additional distinct species have been demonstrated to occur in different parts of the world.

LEFT | The leaf tip of *Hedwigia emodica* is completely white, as if dipped in paint.

DISTRIBUTION
Widespread

ETYMOLOGY
After Johann Hedwig (1730–1799), German bryologist and physician

NUMBER OF SPECIES
12 accepted species

APPEARANCE
Large plants, forming turfs or cushions with irregular branching. Leaves ¹/₁₆–¹/₈ in (1.5–3 mm) long, appressed to stem with leaf tips reflexed when dry, spreading when moist, leaf tip often white, costa absent, leaf cells sinuose and papillose. Monoicous. Perichaetial leaves occasionally with long marginal cilia. Seta short. Capsule immersed in perichaetial leaves, peristome absent. Calyptra small, bell-shaped

ABOVE | These dry shoots of *Hedwigia stellata* show the lower leaf appressed to the stem with the white tip strongly reflexed.

RIGHT | A perichaetial leaf of *Hedwigia stellata* showing the long marginal cilia distinctive of the genus.

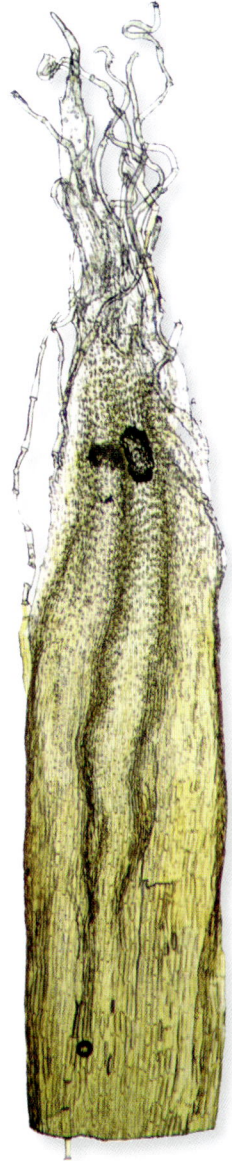

HABITAT
On rocks, rarely epiphytic

SPLACHNUM

The Splachnaceae family includes several genera that are unique among bryophytes for their close relationship with flies. These mosses grow on very specific organic matter such as animal dung, old bones, and decayed carcasses, earning them their common name of Dung Mosses.

Their sporophytes are highly adapted with brightly colored capsules, usually with an expanded base. They also release chemical attractants that lure flies to the capsules. The sticky spores are picked up by the unwitting flies and carried to fresh dung or corpses.

These mosses appear to be restricted to particular substrates. *S. ampullaceum* is found on herbivore droppings, while *Tetraplodon mnioides* only grows on carnivore dung. Laboratory experiments suggest it may be the flies that are the fussy ones, delivering spores to their preferential dung type. A nineteenth-century herbarium specimen of *Tayloria octoblepharum* from Tasmania is particularly notable for its substrate. The specimen label tells us this moss was found "growing on the bones & decayed clothing of a Bushranger, at the base of the Western Mountains, with two double-barrelled Guns & Pistols lying by his side."

The capsules of *Splachnum* plants, especially *S. luteum* and *S. rubrum*, may be the most flamboyant of all the Splachnaceae, the lower part of which in some species is expanded like a wide skirt, giving a superficially flower-like appearance. The success of these mosses is closely linked to their insect partners, and widely reported declines in insect numbers will also affect the long-term survival of the Dung Mosses.

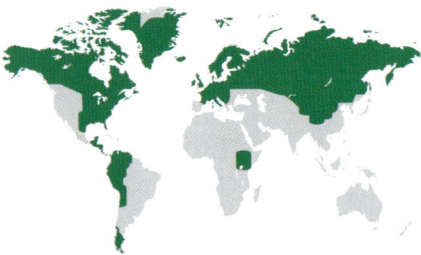

DISTRIBUTION
Widespread mostly in the northern hemisphere

ETYMOLOGY
Unclear, may derive from an Ancient Greek name for a now unknown plant

NUMBER OF SPECIES
11 species

APPEARANCE
Tufts up to 1⅜ in (3.5 cm) tall. Leaves oval to lanceolate, margins smooth or serrate, costa single, leaf cells thin walled and rhomboidal. Most dioicous. Seta long, up to 6 in (15 cm). Capsules with base usually inflated or umbrella-shaped, brightly colored, peristome with 16 teeth

LEFT | The capsules of *Splachnum rubrum* have an expanded base to the capsule that fans out like a skirt and turns a beautiful wine red when mature.

OPPOSITE | In *Splachnum vasculosum*, photographed here in Iceland, the base of the capsule is inflated and reaches a dark purple when mature.

RIGHT | *Splachnum luteum* is recognized by its pale-yellow capsules with a wide "skirt" at the base, growing here in moorland frequented by large herbivores.

HABITAT
On animal dung, found in nutrient-poor environments such as bogs, moors, and tundras

PALUDELLA

The Splachnales includes two families: the Splachnaceae and the Meesiaceae, to which *Paludella* belongs. *Paludella* is a distinctively beautiful moss with strongly recurved leaves arranged in five neat rows up the stem. If you are lucky enough to encounter this plant in its typical fenland or wet meadow habitat, then its identification is unmistakable.

Its one species, *P. squarrosa*, is widespread across the northern boreal and Arctic wetlands. To the south of its range its populations are more fragmented and have significantly declined, particularly in lowland areas, due to the loss and degradation of its wetland habitats. *Paludella* was known from Ireland only as a partially fossilized "subfossil" in Pleistocene peat deposits until an exciting discovery in 1998. Several small populations were found at a fenland site, significantly extending the known range of *Paludella* in Europe.

Evidence of subfossils in peat confirm that *Paludella* was also present across northern England during the late Pleistocene period. It is likely that this striking plant was a common mire moss at the time

LEFT | A close-up view of the *Paludella* shoots shows the dense mat of rhizoids on the lower stem.

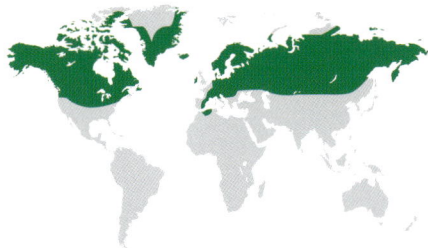

DISTRIBUTION
Restricted to the northern hemisphere, where it is widespread though rare

ETYMOLOGY
Latin *palus* = "marsh" + *ella* = "a diminutive form," alluding to the plant's habitat

NUMBER OF SPECIES
1 species

APPEARANCE
Plants pale green, forming tall turfs up to 6 in (15 cm) high, with dense mat of orange-brown rhizoids on lower stem. Leaves strongly squarrose-recurved, margins finely serrate, costa single, upper leaf cells rounded-hexagonal with single papilla. Dioicous. Seta elongate. Capsules ellipsoid with double peristome

HABITAT
In calcareous fens

when Woolly Mammoths roamed northern Europe. It is speculated that *Paludella's* rich fen habitats were impacted by the changing climate following the end of the ice age, when species associated with these fens became much rarer. However, relics of *Paludella* persisted in northern England until relatively recent times, when its remaining fen habitats were damaged by human disturbance, particularly drainage. The last populations in Britain were seen in the late 1800s, and the species is now regionally extinct here.

BARTRAMIA

Young capsules of *Bartramia* are pale green and spherical, giving them their common name of the Apple Mosses. The capsules are elevated above the mossy cushion on long setae, becoming brown and ridged as they mature. Even when sporophytes are not present, the often blue-green hue of the plants with their long narrow leaves can provide a useful clue to this moss's identity.

The Bartramiales contains one family, the Bartramiaceae, which includes ten genera. The center of diversity for this group of mosses is in the montane tropical region of South America, from where a large number of species have been described.

Bartramia is named after John Bartram, an American botanist who lived in what is now Philadelphia. Bartram was a farmer with a fascination for botany, particularly medicinal plants. He traveled extensively on horseback through the American colonies of eastern North America studying and collecting plants. He corresponded considerably with botanists in Europe. So-called "Bartram's boxes" sailed across the ocean filled with plants that were eagerly received by European gardeners.

BELOW | *Bartramia halleriana* has the typical globe-shaped capsules of the genus, which in this species are held just above the leaves.

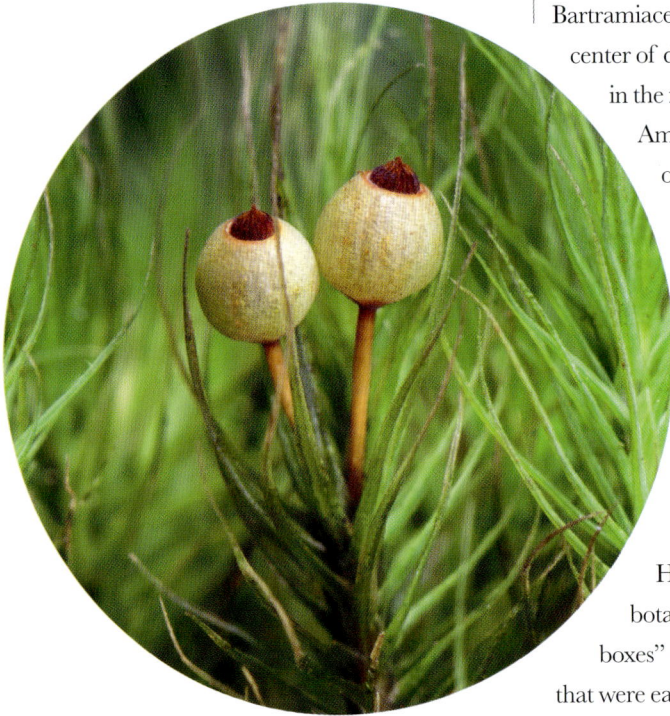

DISTRIBUTION
Widespread, predominantly tropical

ETYMOLOGY
After John Bartram (1699–1777), American botanist and horticulturalist

NUMBER OF SPECIES
57 accepted species

APPEARANCE
Plants forming tufts up to 6 in (15 cm) tall, often densely tomentose. Leaves narrowly lanceolate with sheathing base, leaf blade thickened at margins, margins smooth to toothed, costa single, leaf cells papillose. Dioicous. Seta elongate, rarely short. Capsule rounded to pear-shaped, furrowed when dry, peristome double, single, or absent

HABITAT
On soil or rocks in humid sites

In 1765 Bartram's contribution to botanical exploration was recognized when he was appointed King's Botanist for North America by King George III of Great Britain and Ireland. Bartram showed little interest in mosses, perhaps busy enough with his other botanical pursuits! Some generations later, though, Edwin B. Bartram (1878–1964) took up his famous ancestor's botanical interests with a particular focus on bryophytes and made a major contribution to bryology.

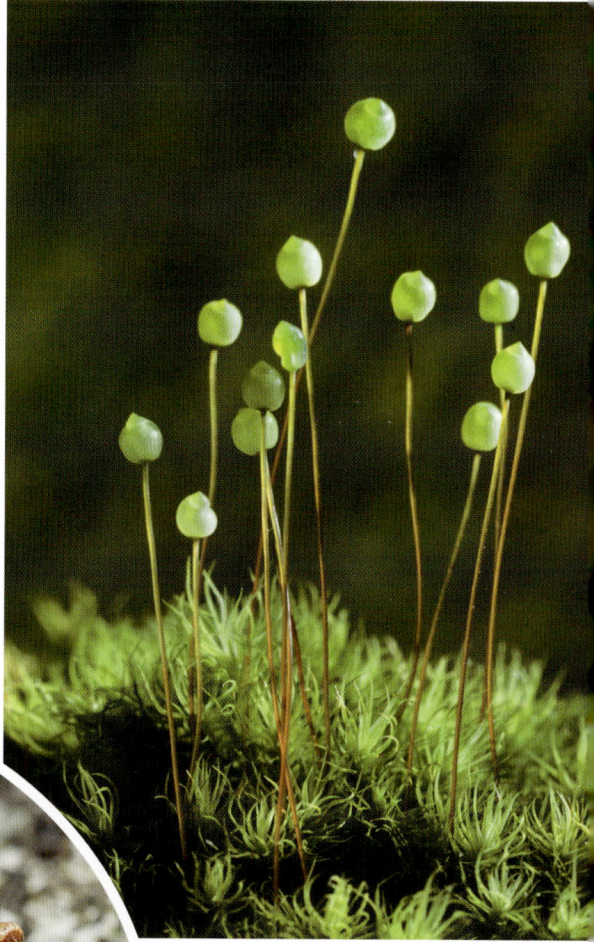

BELOW | The long, silky leaves of this *Bartramia ithyphylla* are a good indicator for the genus, and in this species they will also have a silvery sheathing base.

ABOVE | Capsules of *Bartramia pomiformis* looking very much like miniature green apples at the end of their long, delicate setae.

BRYUM

The *Bryum* genus is a moss that most people could delight in admiring within just a short walk from home. The genus is widespread and some of its species are particularly tolerant of urban environments. *B. argenteum* may very possibly be the most ubiquitous moss on the planet, its tiny silvery cushions filling pavement cracks in cities around the world.

Up until around the late 1990s, *Bryum* was circumscribed as the most species-rich genus of mosses with over 1,000 species described. The classification of the family Bryaceae, and of *Bryum*

DISTRIBUTION
Worldwide

ETYMOLOGY
Greek *bryon* = an ancient word for "moss" or "lichen"

NUMBER OF SPECIES
265 accepted species

APPEARANCE
Plants in various shades of green, brown, and silver, forming cushions usually less than ⅜ in (1 cm) tall. Leaves lanceolate to ovate, costa single, leaf cells quadrate to elongate-hexagonal, thin walled. Dioicous. Seta elongate. Capsule inclined to pendant, peristome double. Vegetative propagation common by specialized bulbils

in particular, has undergone significant changes since then. The problem with *Bryum* was not so much that its classification was based on patterns in morphological similarity but that, for this group of mosses, the morphology provided completely misleading evidence of the plant's evolutionary history. The advent of DNA analysis began to show that groups of species previously thought of as very closely related actually had closer links to species that looked quite different. This has led to the rather unusual situation where the current classification of genera in the Bryaceae family is almost entirely based on molecular DNA data and some genera are no longer defined by morphological features. The current generic classification is based largely on the European flora where species are well studied and molecular data for these species was more readily available. As research continues to look at the Bryaceae in other regions of the world, this classification may likely change again.

OPPOSITE | The silvery cushions of *Bryum argenteum* can be found in a variety of urban environments, often mixed with other acrocarpous mosses, as shown here.

TOP | Leafy shoots of *Bryum argenteum* have a rounded look, like catkins, since the leaves are closely appressed to the stem.

RIGHT | The large, nodding capsules of *Bryum argenteum* are typical of *Bryum* and its closely related genera.

HABITAT
On a variety of substrates such as soil or rocks, less common as epiphytes

RHODOBRYUM

BELOW | A leaf from the upper stem of *Rhodobryum ontariense* that is widest above the middle and ending in a sharp point.

The large attractive plants of *Rhodobryum* are always a joy to find. They can usually be recognized by their bright-green upper leaves, which form distinctive rosettes. In some species these rosettes are formed of up to 20 leaves, resulting in a very striking display. The upright leafy shoots have small, scale-like leaves near the base with larger leaves growing on the upper stem, giving the plants a tiny tree-like structure known as a dendroid growth habit. These secondary stems arise from wiry underground stolons that creep along horizontally. *Rhodobryum* can be found in a range of habitats, most often in forests, but can also occur in sheltered spots of grasslands.

Species of *Rhodobryum* have been used in traditional Chinese medicine, where they are made into a herbal tea to treat a range of ailments including minor heart problems. Based on its long history as a medicinal plant, the biochemical properties of *Rhodobryum* have been analyzed, revealing a range of interesting compounds, including 13 essential oils. Clinical research on extracts of *R. giganteum* has actually found a positive effect in lowering blood pressure. Researchers in 2012 tested extracts of *R. ontariense* for antimicrobial activity and found promising inhibitory activity against specific fungal strains, which could form the

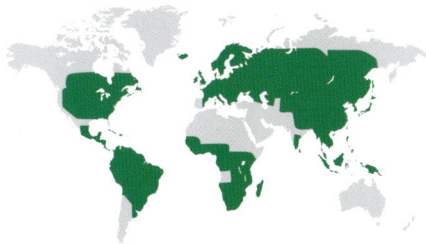

DISTRIBUTION
Widespread

ETYMOLOGY
Greek *rhodon* = "rose" + *bryon* = ancient word for "moss," referring to the terminal leaf rosette in most species

NUMBER OF SPECIES
19 accepted species

APPEARANCE
Plants large, around ⅜–2⅜ in (1–6 cm) tall, often tomentose. Leaves forming distinct rosettes, leaves ovate to spathulate, toothed at margin, leaf cells thin walled and rhomboidal with border of elongate cells, costa single. Dioicous. Seta elongate. Capsule with double peristome

ABOVE | *Rhodobryum ontariense* forms beautiful rosettes of many large, spiraled leaves.

BELOW | This *Rhodobryum roseum* typically has fewer leaves in its terminal rosette than *R. ontariense*.

basis of future antifungal products. These initial findings suggest that bryophytes such as *Rhodobryum* offer a lot of interesting potential in the search for novel pharmaceutical properties.

HABITAT
On soil or occasionally rotten logs and rocks

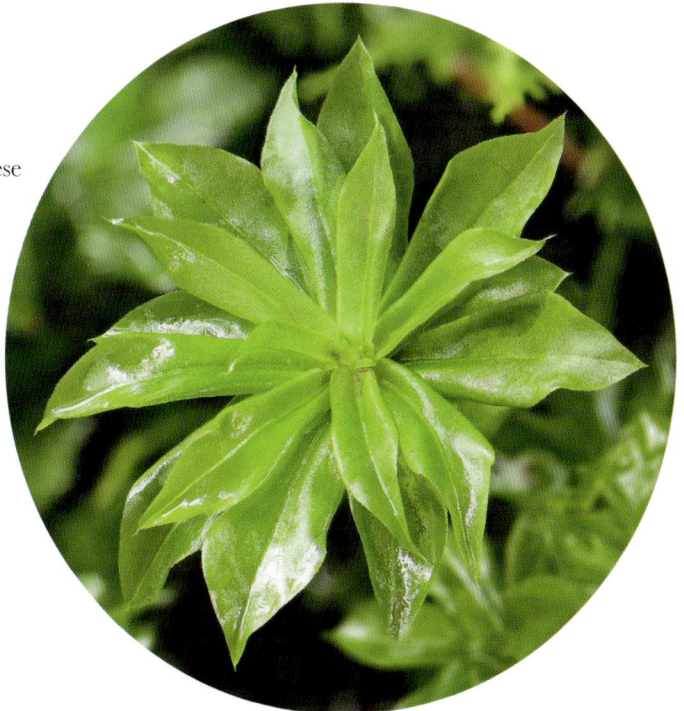

ZYGODON

The Orthotrichales are a clearly defined order represented by one family. The Orthotrichaceae includes 24 genera at the present time, though research into evolutionary relationships is continuing to find more diversity that should be classified at the generic level. The Orthotrichaceae is characterized by the small, roughly rounded upper leaf cells that are often ornamented by papillae and elongated basal cells. The capsules are often ribbed and have a large calyptra, which can be hood- or cone-shaped. The peristome is particularly significant in defining this family, though the double ring of teeth can be variously modified and reduced across the genera.

Zygodon can be most easily recognized in dry conditions when the leaves are appressed to the stem with a characteristic slight twist. Microscopically, the leaf cells are ornamented by several tiny, rounded papillae that help distinguish specimens of *Zygodon* from the superficially similar Pottiaceae, in which the leaf cell papillae are branched or C-shaped. *Zygodon* species produce specialized vegetative propagules called gemmae. These vary between species, and the shape, number of cells, and color of these are important microscopic characters for identifying species. Critical revision will likely reduce the number of accepted *Zygodon* species due to synonymy. However, new species are still being discovered and recorded. The species *Z. catarinoi*, for example, was described only in 2006 from the bryologically well-studied region of the Iberian Peninsula, showing there can still be much to unearth in the field of bryology!

LEFT | *Zygodon viridissimus* is found in Europe and North America. *Zygodon* species can be recognized in the field but species identification usually requires a microscope.

DISTRIBUTION
Widespread, most diverse in tropical regions

ETYMOLOGY
Greek *zygos* = "yoke" (a frame used to join cattle together) + *odon* = "tooth," referring to the paired peristome teeth

NUMBER OF SPECIES
Around 82 accepted species

APPEARANCE
Plants forming loose tufts. Leaves lanceolate, $\frac{1}{16}$–$\frac{1}{8}$ in (1–3.5 mm) long, costa single, upper cells rounded with several papillae per cell. Dioicous or monoicous. Capsule strongly ribbed, peristome double, single, or absent. Calyptra hood-shaped. Gemmae commonly produced

HABITAT
Epiphytic or growing on rocks

LEFT | *Zygodon* species typically grow as epiphytes on trees, such as these extensive cushions of *Z. conoideus* growing on a large trunk.

BELOW LEFT | This *Zygodon conoideus* has young sporophytes with hood-shaped calyptrae.

BELOW | *Zygodon* species are best identified by checking the features of the gemmae microscopically; in this *Z. conoideus* the gemmae are long with a single row of cells.

MACROMITRIUM

The genus *Macromitrium* belongs to a sub-group of Orthotrichaceae mosses that have a creeping primary stem and can form mats with numerous bushy branches. These mosses do not fit into either of the more common pleurocarpous or acrocarpous growth forms.

They are instead described as "cladocarpous," meaning the sporophytes are produced on the ends of short lateral branches, allowing the primary stem to keep growing. When they are producing sporophytes, the plants are recognized by the distinctive large calyptrae covering the capsules.

Macromitrium is a diverse genus that nicely illustrates some of the challenges still to be faced in accurately describing the global diversity of bryophytes. It is a challenging group of mosses since there are many species that are differentiated on quite subtle and variable features. In the early 2000s there were over 50 species names described from Africa and Madagascar. A critical review of these names found that many were describing the same common species, and the actual species diversity was reduced in half.

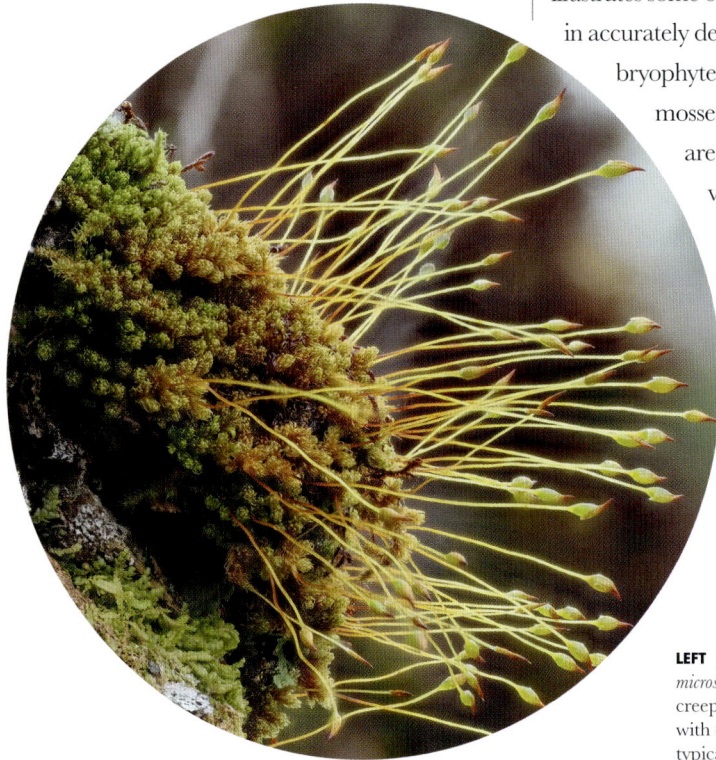

LEFT | This *Macromitrium microstomum* shows the creeping leafy branches with curled dry leaves typical of the genus.

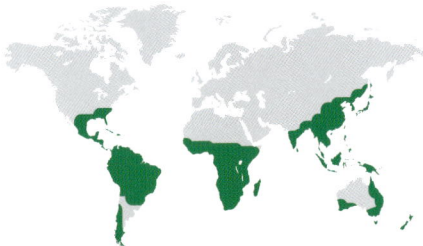

DISTRIBUTION
Widespread across the tropics and southern hemisphere

ETYMOLOGY
Greek *makrós* = "long" + *mitra* = "hat," alluding to the long calyptra

NUMBER OF SPECIES
Around 288 accepted species

APPEARANCE
Plants forming loose mats. Branch leaves lanceolate to lingulate, $\frac{1}{16}$–$\frac{1}{8}$ in (1.5–4 mm) long, costa single, basal leaf cells elongate, upper cells rounded, papillose. Monoicous or dioicous, dwarf males occasional. Seta elongate. Capsule with double peristome. Calyptra large, often hairy

HABITAT
Mostly epiphytic but occasionally on rocks or logs

LEIOMITRIUM

The single species in the *Leiomitrium* genus is a pretty little moss that is distinguished by its creeping branches, its rounded branch leaves, and leaf cells with mostly two or three papillae per cell. There is only one species— *L. plicatum*—and it is only known from just a few localities in the Indian Ocean islands.

L. plicatum has a very restricted distribution and more data is needed to confirm its conservation status. It was first described from a plant discovered on Mauritius over 200 years ago and it has not been recorded since from this island. Its population stronghold is on the island of Réunion. This tiny volcanic island is found 430 mi (700 km) east of Madagascar, and despite its small size it manages to include a wide array of habitats, ranging from a mountainous interior shrouded in montane cloud forest to coastal *Pandanus* groves. *Leiomitrium* has been recorded from several sites around the island but seems most abundant in lowland areas near the coast, which are most at risk of urban development.

ABOVE | *Leiomitrium plicatum* is an attractive moss with creeping branches and leaves with rounded apices.

DISTRIBUTION
Known from Réunion and the Comoros

ETYMOLOGY
Greek *leios* = "smooth" + *mitrium* = "hat," presumably alluding to the non-plicate calyptra

NUMBER OF SPECIES
Only 1 species

APPEARANCE
Plants creeping, branches 2–3 in (5–8 cm) long. Leaves slightly flexuose in dry state, stem leaves lanceolate, branch leaves with obtuse apex, costa single, upper leaf cells rounded, usually with 2–3 papillae per cell. Dioicous. Seta stout. Capsule strongly ribbed with double peristome. Calyptra cone-shaped, hairy

HABITAT
Growing on trees or rocks

ORTHODONTIUM

The Orthodontiaceae are a small family that has sometimes been placed in the Rhizogoniales but is now recognized at ordinal level. *Orthodontium* mosses form dense turfs with long, narrow leaves and are typically encountered on rotting tree stumps in woodland.

In Europe the native *O. gracile* is a rare species in decline. More common here is the southern-hemisphere species *O. lineare*, which was introduced to Europe in the early 1900s. *O. lineare* has potentially contributed to the decline of *O. gracile*, since it is a much stronger competitor. These two species are actually very hard to distinguish from one another since they look so similar, though research published in 2019 has shown that, in terms of their evolutionary history, these two mosses are surprisingly different. These two species lineages diverged over 50 MYA toward the end of the reign of the dinosaurs. *O. gracile* is more closely related to *O. lignicola*, a species from the Sino-Himalaya, from which it actually looks rather different. This research illustrates how looks can be deceiving when thinking about a plant's closest relatives.

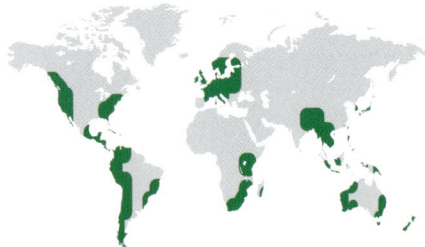

BELOW | *Orthodontium lineare* with its long, thin leaves and long, cylindrical capsules.

DISTRIBUTION
Worldwide, predominantly in tropical and temperate southern-hemisphere regions

ETYMOLOGY
Greek *orthos* = "upright" + *odontos* = "tooth," referring to the upright peristome teeth

NUMBER OF SPECIES
9 accepted species

APPEARANCE
Forming tufts less than ⅜ in (1 cm) high. Leaves sometimes curved in one direction, oblong to linear-lanceolate, apex minutely toothed, costa single and narrow, leaf cells elongate and smooth. Monoicous. Capsule cylindrical with tapering neck, peristome double

ABOVE | *Orthodontium lineare* is usually found on very rotten, crumbling tree stumps and logs in woodland.

HABITAT
On rotting wood, occasionally on soil and base of trees

O. gracile also occurs in North America, where it hybridizes with another species found there: *O. pellucens*. Hybrids produced from these species have the appearance of *O. gracile* but the chloroplast genes of *O. pellucens*. Understanding the complex genetics and evolutionary history of mosses can be an important tool for prioritizing conservation actions, in this case lending a greater significance to European *O. gracile* populations.

PYRRHOBRYUM

The Rhizogoniales are an early diverging lineage of pleurocarpous mosses. Pleurocarpy in mosses is when the females sex organs are produced on very short branches that are specialized for this purpose, having only perichaetial leaves and no vegetative leaves. This innovation has freed the pleurocarpous mosses from the more restrictive growth form of

BELOW | The long feathery branches of *Pyrrhobryum dozyanum*, photographed here in southern Japan.

DISTRIBUTION
Widespread across tropical and southern temperate regions

ETYMOLOGY
Greek *pyrrhos* = "fire-colored" + "bryon" = an ancient word for "moss," possibly referring to the color of the peristome

NUMBER OF SPECIES
9 accepted names

APPEARANCE
Plants forming tufts up to 2 in (6 cm) tall. Leaves twisted when dry, linear-lanceolate to oblong, leaf margin with paired teeth, costa single, leaf cells small, quadrate. Dioicous or monoicous. Capsules with double peristome

acrocarpy, where the production of the sex organs terminates the growth of the branch it develops on. The Rhizogoniales lineage is partially pleurocarpous and does not show all the distinctive features of the more derived pleurocarps. In fact, their growth form can resemble the acrocarpous mosses since they tend to grow in loose tufts that branch from the very base of the primary stems. The Rhizogoniales tend to grow on soil or dead wood and their geographic range is predominately southern hemisphere.

The feathery, dark-green shoots of *Pyrrhobryum*, often tomentose with dense red-brown rhizoids below, form large clumps on the forest floor and are commonly encountered in tropical regions. A closer look under the magnifying glass reveals rows of sharp teeth, not only along the leaf margins but along the back of the leaf costa as well.

RIGHT | *Pyrrhobryum* species often grow as cushions on the floor of tropical forests, as shown in this example from Taiwan.

HABITAT
On soil, rocks, or tree bark in moist sites

AULACOMNIUM

BELOW | The widespread *Aulacomnium androgynum* requires a slightly acidic substrate, growing mostly on decaying organic matter like this rotting tree stump.

BELOW | The widespread *Aulacomnium androgynum* requires a slightly acidic substrate, growing mostly on decaying organic matter like this rotting tree stump.

The Aulacomniaceae family is a group of mosses whose nearest relatives have remained ambiguous. They are sometimes regarded as part of the Rhizogoniales, but a more established consensus treats them at ordinal level. Like the Rhizogoniales, they are considered to be an early diverging lineage of pleurocarpous mosses. *Aulacomnium* mosses can be variable in their features, for example in their leaf shape, and a worldwide revision is needed to confirm the number of distinct species, which will probably be found to be around four.

Some species of *Aulacomnium* can reproduce vegetatively by producing gemmae. Shoots of *A. androgynum* often end in a slender stalk that has a ball of gemmae at the end. These specialized gemmae stalks are quite rare in mosses and are called a "pseudopodia." The gemmae balls are formed of a cluster of elongate, pale-green gemmae and the whole structure is reminiscent of a tiny green drumstick.

The Arctic moss *A. turgidum* has been found to be incredibly tolerant of freezing stress. In a remarkable show of survival skills, it was one of the species that was observed regrowing after 400 years of being entombed in ice following the melting of a Canadian glacier. *A. turgidum* is most at home in the windswept,

DISTRIBUTION
Widespread in temperate regions and tropical mountains

ETYMOLOGY
Greek *aulacos* = "furrow" + *Mnium*, suggesting this genus is similar to the *Mnium* mosses but with a furrowed capsule

NUMBER OF SPECIES
6 accepted species

APPEARANCE
Dense yellow-green tufts, up to 2 in (6 cm) tall, usually tomentose below. Leaves ovate to lanceolate, margins often minutely toothed, leaf cells quadrate, usually papillose. Dioicous. Seta elongate. Capsule furrowed, peristome double. Gemmae occasional on stalks

BELOW | *Aulacomnium turgidum* is a moss of cold extremes, mostly restricted to Arctic and alpine regions.

ABOVE | The balls of clustered gemmae produced by these *Aulacomnium androgynum* plants are raised up on stalks called pseudopodia.

frozen Arctic tundra. Fossil evidence suggests this species was once much more common south of its present range in eastern North America when the last ice age expanded the range of tundra vegetation.

HABITAT
On soil, rotting wood, or rocks; occasionally epiphytic

RACOPILUM

BELOW | *Racopilum* cf. *capense*, in the montane tropical cloud forest of Réunion Island.

The Hypnodendrales are an order of pleurocarpous mosses. The pleurocarps appeared relatively late in the evolution of mosses, with research suggesting an origin in the Late Jurassic. Around the same point in geological time, the rise of angiosperm forests may have provided an increase in potential habitat niches, triggering an expansion in their speciation. The Hypnodendrales were the first order to diverge from this pleurocarpous moss lineage and so hold interesting clues about the origins of this successful group. Mosses in this order are largely restricted to the southern hemisphere.

Racopilum is a tropical and subtropical genus of creeping mosses with strongly dimorphic leaves. This leaf arrangement is particularly unusual for mosses. The shoots are rather flattened and have two rows of larger lateral leaves and two rows of small dorsal leaves on the upper side with rhizoids on the underside. A high number of species names are accepted but a critical review of these will likely find a much lower number of distinct species.

DISTRIBUTION
Widespread across tropical regions and the southern hemisphere

ETYMOLOGY
Greek *rhakos* = "ragged" + *pilos* = "felt hat," referring to the frayed calyptra

NUMBER OF SPECIES
44 accepted species

APPEARANCE
Tangled mats, often tomentose. Leaves dimorphic with dorsal leaves small, and lateral leaves larger, oblong to lanceolate, $\frac{1}{16}$ in (1–2 mm) long, with costa excurrent and leaf cells rounded to hexagonal. Monoicous or dioicous. Seta elongate. Capsules with double peristome. Calyptra hairy

HABITAT
On trees, soil, or rocks

SPIRIDENS

When considering what might be the largest and most magnificent moss of all, *Spiridens* would certainly be a close contender. It grows in tropical forests, climbing up tree trunks rather like a vine with its long sturdy branches fanning out from the tree. *S. reinwardtii* can climb to incredible heights of up to 10 ft (3 m) up a tree trunk—truly one of the wonders of the moss world.

S. muelleri is an endemic species on Lord Howe Island, but somewhat unexpectedly given its distribution across the Southeast Asian tropics, *Spiridens* has not yet been found to occur in the Australian wet tropics.

Spiridens has been used traditionally in some countries such as the Philippines and Malaysia to decorate headdresses and clothing as well as ward off evil spirits. In the Philippines the long strands of this moss are apparently still used as a binding material.

RIGHT | The magnificent *Spiridens reinwardtii* climbing up a tree trunk in the tropical forest of Sabah, Malaysia.

DISTRIBUTION
Found across Southeast Asia and Lord Howe Island, Australia

ETYMOLOGY
Latin *spira* = "coiled" + *dens* = "tooth," referring to the peristome, which can be spiraled when dry

NUMBER OF SPECIES
9 species

APPEARANCE
Plants very large, branches to 20 in (50 cm) long. Leaves lanceolate with base clasping stem, differentiated leaf cells forming a marginal border, costa often extending into short awn, Dioicous. Seta short. Capsule with peristome spirally inrolled

HABITAT
Epiphytic, typically growing on tree trunks in tropical forests

211

PTYCHOMNION

The Ptychomniales is a fairly small order of pleurocarpous mosses comprising three families, the largest being the Ptychomniaceae with 12 genera. *Ptychomnion* is a genus of large, attractive mosses characterized by their squarrose, rugose leaves, and with the plants growing in turf-like colonies. In some regions it is colloquially referred to as the Pipe-Cleaner Moss. Its long stems with their crumpled bushy leaves have a rather papery texture. Whether or not the leaf is attached to the stem with a sheathing base is a useful distinction for different *Ptychomnion* species. Most species have longitudinal folds in the leaves, though some plants show these more distinctly than others.

Ptychomnion is commonly found in temperate wet forests of the southern hemisphere, most frequently in South America, southeastern Australia, and New Zealand. Although *Ptychomnion* produces separate male and female plants, the males are minuscule dwarf plants that perch up upon the leaves of the female. This clever adaptation resolves the problem of proximity that many dioicous species face.

LEFT | This *Ptychomnion aciculare* is part of the bryophyte flora of New Zealand's wet forests.

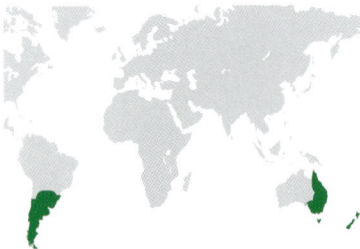

DISTRIBUTION
Found in temperate to sub-Antarctic southern hemisphere regions

ETYMOLOGY
Greek *ptychos* = "pleated" + *mnion* = an old word for "moss," referring to the pleated the leaves

NUMBER OF SPECIES
4 accepted species

APPEARANCE
Plants sparsely branched, up to 6 in (15 cm) long. Leaves squarrose, wrinkled, ovate, costa absent or short, leaf cells linear. Dioicous, sometimes producing dwarf males. Seta black. Capsule inclined, with double peristome

LEFT | *Ptychomnion aciculare* has squarrose leaves that are minutely crumpled, giving the stems a distinct look.

RIGHT | The tiny dwarf male plants of *Ptychomnion aciculare* are visible here as a cluster of leaves at the tip of the female plant leaf.

HABITAT
On soil or rotting logs, or epiphytic

HYPOPTERYGIUM

The leafy green umbrellas of *Hypopterygium* resemble a forest of tiny palms. It is one of several bryophyte genera that display this tree-like, or dendroid, growth habit. Its wiry stems grow horizontally, giving rise to upright secondary stems bearing a flattened fan of leafy fronds. The leaves of these fronds have distinctly dimorphic leaves that are usually differentiated into large asymmetric lateral leaves and a single row of small leaves on the underside of the stem. This leaf arrangement is unusual for mosses, which typically have leaves spiraled around their stems, and is more characteristic of the leafy liverworts.

The Hypoterygiaceae family includes eight genera, and its relationships within the pleurocarpous mosses have previously been uncertain. Recent molecular work has confirmed that the Hypoterygiaceae form a distinct lineage best recognized at the ordinal rank as a sister group to the large orders of the Hookeriales and Hypnales.

Hypopterygium has a pantropical and southern temperate distribution with some extensions into northwest America and Japan. The most widespread species, *H. tamarisci*, has been introduced to Europe, where it can be found in greenhouses in Scotland (UK), Germany, and Switzerland. The only European locations where it has been reported to be established outside are arboretums in Portugal and northern Italy. Work by scientists in the year 2000 analyzed the DNA signatures of plants collected from the European locations and concluded that these plants were most closely related to Australian populations and were probably imported with tree ferns or exotic trees.

LEFT | A flattened frond of *Hypopterygium flavolimbatum*, in British Columbia, Canada.

DISTRIBUTION
South hemisphere, also Pacific northwest America, and introduced in Europe

ETYMOLOGY
Greek *hypo* = "under" + *pterygion* = "little wing," referring to the small underleaves

NUMBER OF SPECIES
12 accepted species

APPEARANCE
Plants dendroid, up to ¾ in (2 cm) tall. Leaves with margins bordered by rows of elongate cells, inner leaf cells roughly hexagonal, costa single. Dioicous or monoicous. Capsules pendulous, peristome double. Calyptra hood-shaped

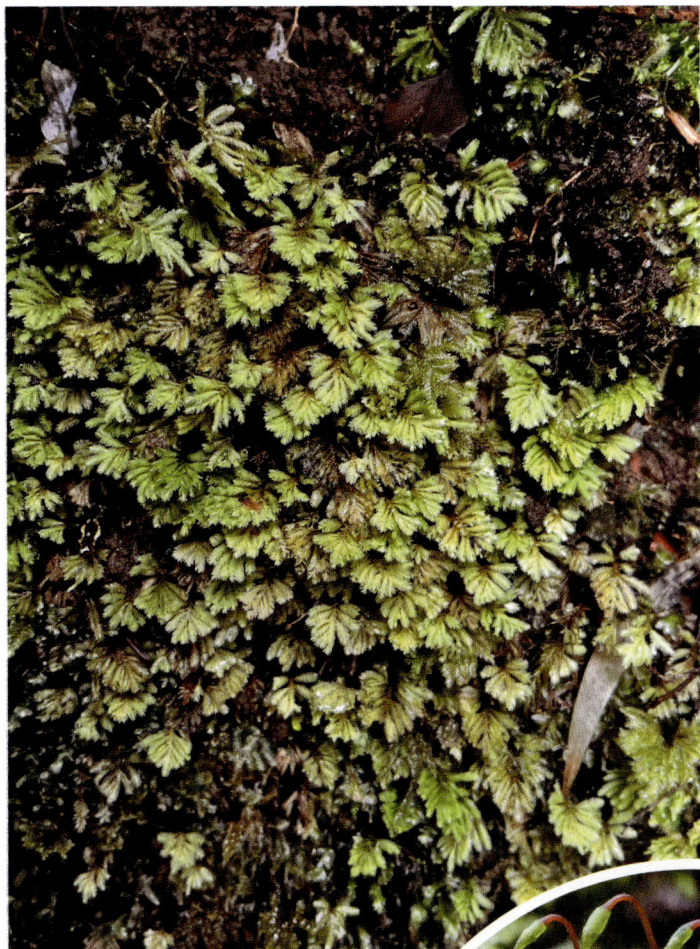

ABOVE | A forest of *Hypopterygium didictyon* ascending up a tree trunk in a Chilean rainforest.

RIGHT | *Hypopterygium tamarisci* plants in the montane forests of Réunion Island with tall upright sporophytes.

HABITAT
Usually growing on the forest floor, on tree trunks or logs

DALTONIA

The Hookeriales is a large order of tropical and subtropical mosses. This order, along with the Ptychomniales and Hypnales, form a group of related mosses that includes an estimated 98 percent of pleurocarpous moss species. The Hookeriales are distinguished in part by their thin-walled leaf cells that tend to be short rather than elongated and the lack of differentiated alar cells at the leaf base.

Daltonia is predominantly tropical and is found throughout the tropics, mostly in mountainous areas. The small, sparsely branched tufts are typically found on twigs. The plants have small, lanceolate-shaped leaves with a distinct border of elongate cells along the leaf margin and a single, well-defined costa. The plants are often found with sporophytes and are then easily recognized by the pale calyptra that is densely fringed at the base. William Jackson

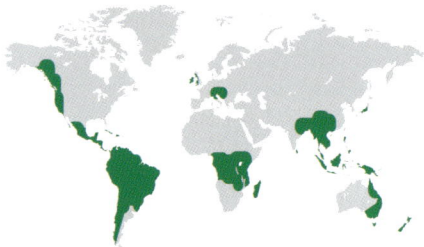

DISTRIBUTION
Predominately tropical, with some extensions into temperate regions

ETYMOLOGY
After James Dalton (1764–1843), English clergyman and botanist

NUMBER OF SPECIES
30 accepted species names

APPEARANCE
Plants forming glossy, pale-green tufts up to ¾ in (2 cm) tall. Leaves lanceolate, upper leaf smooth or minutely toothed, leaf cells oval-shaped with a distinct border of linear cells at the margin. Monoicous. Seta elongate. Capsule urn-shaped, peristome double. Calyptra cone-shaped, fringed at base

HABITAT
Usually epiphytic, rarely on rocks

OPPOSITE | *Daltonia splachnoides* with its characteristic capsules with frilled calyptrae, growing among the liverwort *Metzgeria*.

ABOVE | *Daltonia* usually grows as small tufts on trees and shrubs, such as this *Daltonia splachnoides* perched on a tree trunk.

BELOW | A leaf of *Daltonia splachnoides* as seen under the microscope with clear angular leaf cells bordered by elongate cells at the margin.

Hooker named *Daltonia* after his close friend and fellow botanist James Dalton, whom he also made godfather to his son Joseph Dalton Hooker.

D. splachnoides has a globally disjunct distribution pattern where it is common in tropical regions with isolated occurrences in the temperate northern regions of northwestern America, Ireland, Great Britain, and Madeira. It is reliant on high rainfall and tends to be rather rare in these areas. Interestingly, although *D. splachnoides* is still a rare moss in Britain, it has been increasing its range. It is not clear what has caused this shift, but it could possibly be due to changes in climate.

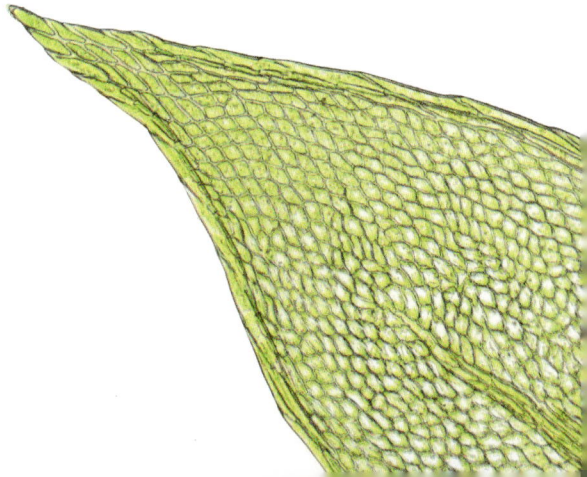

HOOKERIA

The handsome mosses of *Hookeria* are a pale glossy green with large, soft leaves that appear quite translucent when moist. The ovate or lanceolate leaves are flattened along the stem and the plants can at first glance be mistaken for a large leafy liverwort. The leaf cells are so big they can be seen with a ×10 magnifying glass. The sporophytes typically have pendant or inclined capsules and a double peristome with exostome teeth and the endostome a high basal membrane. The large, thin-walled leaf cells, smooth leaf margins, and lack of a costa in the leaf help to distinguish *Hookeria* from other pleurocarpous mosses.

The nineteenth-century concept of *Hookeria* was morphologically much broader and included many species now classified in other families; a total of 506 names have at some point been assigned to *Hookeria* in the last 200 years! It is now defined more narrowly, with 12 species accepted. Some of these names are still poorly known and need further investigation. There are two widespread species in the genus: *H. lucens* and *H. acutifolia*.

LEFT | A luxuriant pale-green patch of *Hookeria lucens* growing on a mossy stream bank.

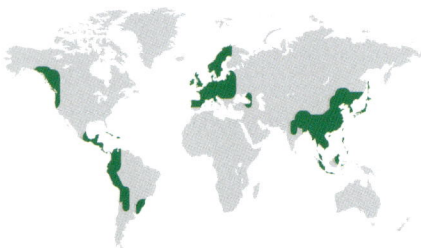

DISTRIBUTION
Pantropical with some extensions into temperate Europe and North America

ETYMOLOGY
After William Jackson Hooker (1785–1865), English botanist and first director of Royal Botanic Gardens, Kew

NUMBER OF SPECIES
12 accepted species

APPEARANCE
Pale, glossy, succulent green mats, stems to around 2 in (5 cm) long. Leaves ovate to broadly lanceolate, leaf cells hexagonal or rhomboidal and very large, costa absent. Monoicous. Seta elongate and stout. Capsule cylindrical, peristome double. Calyptra cone-shaped

HABITAT
On moist soil and rock

The genus name commemorates the famous botanist William J. Hooker, who, along with his son Joseph Dalton Hooker, was highly influential in the development of botanical science at this time. Both father and son had a particular interest in bryophytes and produced many important publications themselves as well as using their extensive networks to connect botanists and further bryological research.

RIGHT | *Hookeria lucens* is a rather large moss that forms soft glossy cushions with shiny leaves.

LEFT | A shoot of *Hookeria acutifolia* with pointed leaf apex, in British Columbia, Canada.

CYCLODICTYON

Species of *Cyclodictyon* occur mostly in the tropics of Central America and tropical Africa. The plants form loose mats with flattened shoots and tend to be found in wet places. The leaves are bordered by rows of elongate cells that contrast with the more rounded-hexagonal, large cells in the rest of the leaf. The long, slender costa is forked and extends up most of the leaf. *Cyclodictyon* probably has far fewer species than the current number of names accepted and is in need of worldwide revision.

C. laetevirens is the only species to extend into the temperate north, where it is rare and has a scattered distribution in hyperoceanic regions that have high levels of rainfall all year round and mild winters. It has a particular stronghold in the Macaronesia islands. In 1840 the botanist William Borrer came across this beautiful plant growing in a sea cave while walking on the Cornish coast of England. On returning home he realized this species was a new record for the region so he asked a local friend to go to the cave and collect a specimen so he could confirm the find. Disastrously, the misguided friend stripped all the *Cyclodictyon* from the cave! In the commotion that followed, Borrer discovered that John Ralfs had discovered the same plant in a sea cave along the coast some months previously. Thankfully, this population persists to this day in Ralfs's cave, its only remaining site in England, where it is protected under UK law.

RIGHT | A leaf of *Cyclodictyon laetevirens* under the microscope showing the long double nerve and border of long narrow cells.

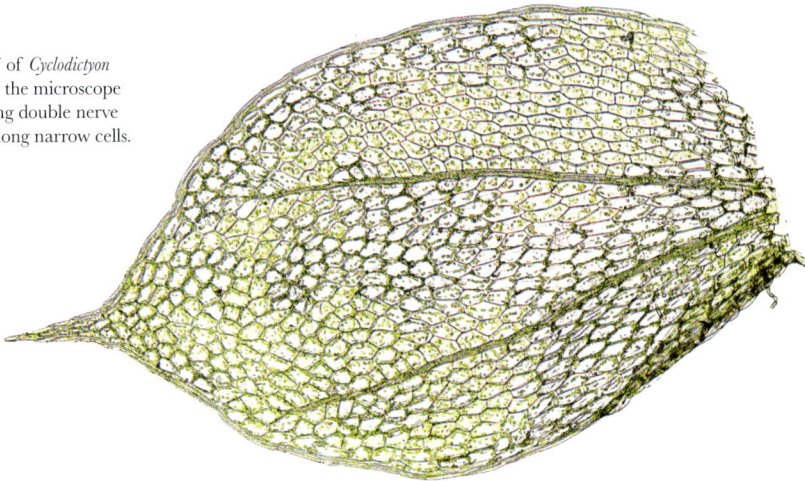

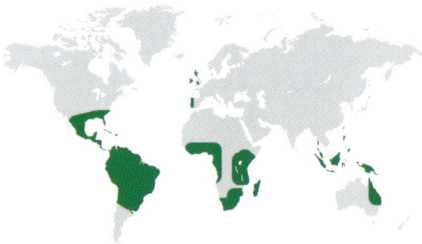

DISTRIBUTION
Widespread across tropical regions, with rare occurrences in Europe

ETYMOLOGY
Greek *kúklos* = "circle" + *diktyon* = "net," referring to the large, rounded leaf cells

NUMBER OF SPECIES
83 accepted species

APPEARANCE
Pale-green mats up to around 3 in (8 cm) long. Lateral and dorsal leaves differentiated, oblong-ovate, leaf cells large, hexagonal, bordered by linear cells at leaf margin, costa double. Monoicous or dioicous. Capsule with double peristome. Calyptra with fringed base

LEFT | The African *Cyclodictyon vallis-gratiae*, shown here covering an extensive area of a cave wall.

BELOW | Creeping branches of *Cyclodictyon laetevirens* showing the broadly oval leaves ending in a small point.

HABITAT
On soil, occasionally on rocks or epiphytic

HYPNUM

BELOW | The species *Hypnum cupressiforme* is a very common member of the genus that can also be quite variable; its leaves are curved and can be an attractive brownish green, as shown here.

OPPOSITE TOP | *Hypnum uncinulatum* is a rare moss of Atlantic Europe, shown here growing over a tree branch in the rainforest of Madeira.

The Hypnales is by far the largest order of mosses, with 469 genera and over 4,000 species—it includes almost a third of all mosses! They can usually be separated from other pleurocarpous mosses by features such as the differentiated alar cells at the base of the leaves and the smooth capsules. Establishing the evolutionary relationships within this large order has been challenging, since they are believed to have diversified relatively recently during the Late Jurassic and Early Cretaceous periods, and the molecular data shows a rather low level of diversity across the different genera. The Hypnales are predominantly distributed in the northern hemisphere.

Plants in the genus *Hypnum* tend to form large, glossy mats with leaves distinctly curled in one direction and often lacking a costa. They have cylindric capsules and a well-developed peristome. *Hypnum* was first officially described by Johannes Hedwig in 1801, based on an earlier description by Johann Dillenius, and the definition of this genus has varied hugely over the following

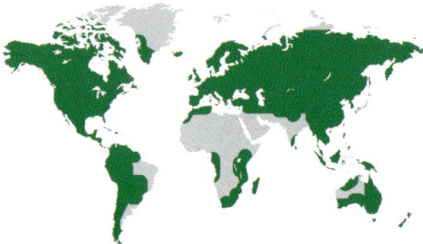

DISTRIBUTION
Worldwide except Antarctica

ETYMOLOGY
Greek *hypnos* = "sleep," referring to the use of this moss as pillow stuffing

NUMBER OF SPECIES
73 accepted species

APPEARANCE
Plants forming branched mats. Leaves straight to strongly curved, lanceolate to ovate, narrowing to a filiform apex, costa short, double or absent, leaf cells linear and smooth, alar cells small and quadrate. Dioicous. Seta elongate. Capsule cylindric, peristome double

years. It once included most of the pleurocarpous mosses, but over the last two centuries it has become more narrowly defined as it has been reclassified into various other distinct genera.

Hypnum is a common moss that often grows in abundance, and it has also been widely used by people through the ages, mostly as stuffing for cushions. It is likely that Dillenius originally chose the name Hypnum because of the moss's association with sleep, given its popularity at the time as a pillow stuffing.

HABITAT
Usually epiphytic on trees, also grows on soil or rocks

BELOW | *Hypnum cupressiforme* var. *lacunosum* has capsules ending in a long narrow beak.

MYURIUM

The *Myurium hochstetteri* species is a beautiful moss forming shimmering golden-green cushions. Its leaves are very concave and they constrict abruptly into a long tail-like point, hence the origin of the genus name and the common name in some regions of Hare-Tail Moss.

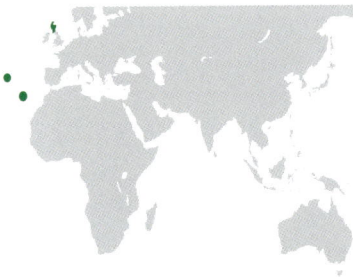

BELOW | In western Scotland, *Myurium hochstetteri* can be found amid rocky coastal sites like this scene on the Hebridean island of Eigg.

DISTRIBUTION
Found in Scotland and the Macaronesia islands

ETYMOLOGY
Greek *myouros* = "mouse-tailed," referring to the appearance of the branches

NUMBER OF SPECIES
1 confirmed species

APPEARANCE
Plants with branched, upright stems, up to around 2 in (6 cm) tall. Leaves very concave, ovate to lanceolate, narrowed into a long apiculate point, costa absent, leaf cells elongate with differentiated alar cells. Dioicous. Setae elongate. Capsules curved, with double peristome. Calyptra hood-shaped

HABITAT
Epiphytic or on rocks

M. hochstetteri has an interesting disjunct distribution. It grows at specific sites in Scotland, mainly around the Hebridean islands, and also occurs in the Macaronesia islands of the Azores, Madeira, and Tenerife.

In Scotland it grows on rocks or grassy banks by the coast. Its rather scattered distribution here is something of a puzzle since there are many sites in this region that have similar conditions to the habitats it does occur in, but from which the plant is inexplicably absent. *Myurium* is much more frequent and widespread in the Azores and Madeira, where it more commonly grows as an epiphyte in forests. It is rather a rare plant in Tenerife. In 1969 it was discovered for the first time from Ireland, but despite thorough searching, it has not been re-found and is presumed to no longer occur here.

There are three additional *Myurium* species names recorded from Asia but the status of these is ambiguous. It is likely that these names refer to plants that should be reclassified in other genera.

ABOVE RIGHT | *Myurium hochstetteri* has ovate, concave leaves that narrow to a long point, as shown in this epiphytic plant from Madeira.

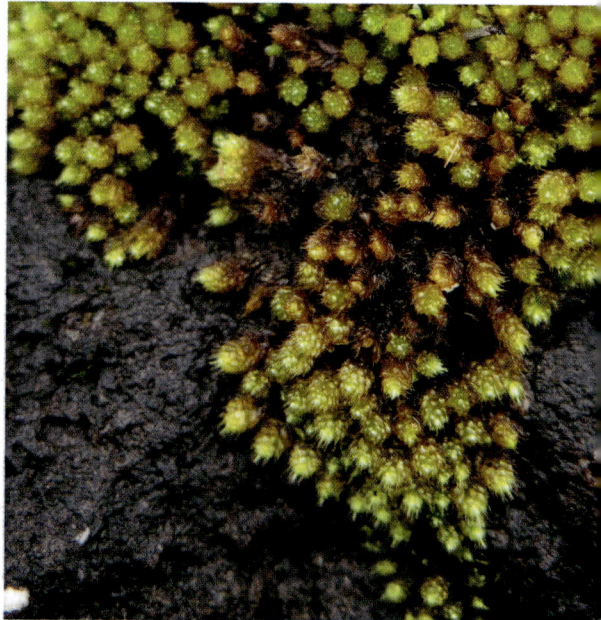

RIGHT | The Scottish *Myurium hochstetteri* glistens against the dark rocks of Eigg.

DREPANOCLADUS

The Amblystegiaceae is a family of pleurocarpous mosses that includes many species adapted to wetland habitats. It is particularly diverse in the northern temperate regions of the world such as North America and Europe. The taxonomy of the Amblystegiaceae has been notoriously difficult. One of the challenges when studying mosses growing in wetlands is that their morphological features can be very variable. In these habitats the plants may be submerged for part of the year and exposed to drier conditions at other times. The plants respond to these differing conditions in their growth. When underwater they may grow larger and some features of the leaf structure may be more strongly developed; conversely, they may grow as smaller plants when perched above the water line. When a series of specimens are then reviewed to define a species, these kinds of variations can cause a lot of confusion.

Drepanocladus typically grows in wet, marshy areas or along streams, though in some places it can also grow completely submerged in lakes. The plants are rather brown in color, and typically leaves are curved over in one direction, giving the plant a swept-back look. Wetland habitats in many regions of the world have been reduced by drainage or degraded. This trend has a direct impact on the wetland species that depend on these habitats. Consequently, a conservation review of all European moss species has found that over half of the *Drepanocladus* species have some level of threat to their continued existence.

LEFT | *Drepanocladus*, like this species *D. sendtneri*, typically have very curved leaves.

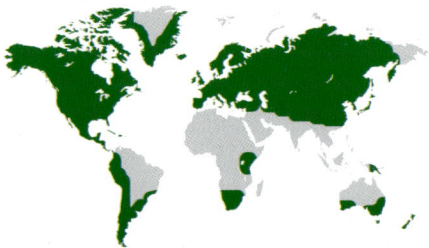

DISTRIBUTION
Nearly worldwide

ETYMOLOGY
Greek *drepane* = "sickle" + *clados* = "branch," referring to the curved branch leaves

NUMBER OF SPECIES
19 accepted species

APPEARANCE
Plants small to large, yellow green, often brownish. Leaves straight to strongly curved, ovate to lanceolate, gradually tapered to a fine point, costa single, alar cells inflated. Monoicous or dioicous. Seta long. Capsule cylindrical, curved, with double peristome

HABITAT
Typically found in nutrient-rich wetlands, ditches, pools, and swampy forests

ABOVE | Brownish plants with hooked shoot tips and curved leaves are a good indicator for *Drepanocladus*, as illustrated by this *D. angustifolius*.

LEFT | The rare European moss *Drepanocladus lycopodioides* has declined due to the drainage and pollution of its aquatic habitats; photographed here in North Wales, UK.

FONTINALIS

There are very few truly aquatic moss genera, of which *Fontinalis* is one. These plants are always found associated with water, though some species may live in seasonally wet habitats, where the plants can survive out of water for part of the year. Unusually for bryophytes, *F. dalecarlica* is also able to tolerate low levels of salinity and grows in the northern Baltic Sea.

Fontinalis leaves are often keeled into a boat shape and usually arranged in three distinct rows down the stem. The creeping branches and indeterminate growth pattern of *Fontinalis*, combined with the buoyancy effect of growing in water, mean that these plants can reach great lengths of well over 30 in (80 cm) in extreme examples. As with other mosses closely associated with water, the setae are very short, leaving the capsules immersed within the leaves. Sporophytes are, however, rarely produced.

Plants of the same species can show a striking variation in their morphology, which is often a feature of aquatic plants. A total of 256 names have been classified under *Fontinalis* over the last two centuries, and these have now been refined down to

LEFT | *Fontinalis antipyretica* can form substantial mats covering river boulders, as shown here at the source of the river Doubs in Western France.

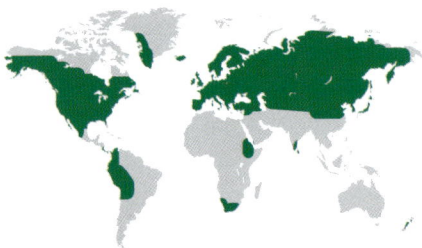

DISTRIBUTION
Widespread across the northern hemisphere, South America, Africa, and Asia

ETYMOLOGY
Latin *fontanus* = "in or by a spring," as usually found growing in running water

NUMBER OF SPECIES
17 accepted species

APPEARANCE
Large plants with trailing stems, reaching up to 30 in (80 cm) long. Leaves range from lanceolate to ovate, keeled, concave, costa absent, cells in midleaf elongate, alar cells enlarged. Dioicous. Seta short. Capsule ovoid to cylindric, peristome double. Calyptra tiny

17 accepted species. The most widespread species is *F. antipyretica*, which was named by Linnaeus in reference to its use by Swedish peasants as a filling between the chimney and walls of a house in an attempt to prevent fires.

HABITAT
Submerged rivers and water bodies

SEMATOPHYLLUM

T he genus of *Sematophyllum* is a large, mostly tropical genus with just a few species occurring in northern temperate regions. Leafy plants of *Sematophyllum* are best defined by the large alar cells that form a row at the leaf base near where the leaf attaches to the stem and by the lack of a costa. When sporophytes are present, the setae are very long and the capsules seem rather small in proportion.

The Sematophyllaceae is one of the largest families in the Hypnales. Many of the accepted species of *Sematophyllum* have not been studied since they were first described, which could be well over 100 years ago, so it is challenging to envisage the true global diversity of this genus. A critical revision of all the world's species would be very helpful, albeit rather a large task.

BELOW | *Sematophyllum dregei*, a plant of southern Africa, showing the curved leaves and inclined capsules typical of *Sematophyllum*.

DISTRIBUTION
Widespread across tropical regions with some temperate extensions

ETYMOLOGY
Greek *sematos* = "signal" or "flag" + *phyllon* = "leaf," referring to the inflated alar cells

NUMBER OF SPECIES
Around 170 accepted species

APPEARANCE
Plants forming glossy golden-green mats. Leaves ovate to lanceolate, costa absent, cells elongate and smooth, alar cells inflated. Monoicous or dioicous. Seta elongate. Capsule usually with double peristome. Calyptra hood-shaped

HABITAT
Epiphytic on trees, also on rocks and occasionally soil

ORTHOSTICHELLA

The *Orthostichella* genus is moss of tropical forests, where it often grows dangling from tree branches in pendant wefts. The plants have a creeping stem with appressed leaves that grows along tree bark, and the branches that grow from this primary stem can be erect or pendant. The branch leaves are usually arranged in spiraled rows around the stem to beautiful effect. The leaf costa can be quite variable in its presence or absence in *Orthostichella*, which is unusual since this is normally quite a consistent feature of a species. Even within a single plant, the leaves can vary from having a single, double, or absent costa.

Mosses only adopt a pendulous growth form in very humid sites such as tropical forests, where the plants can trap humidity and not risk drying out. At the leaf base of *Orthostichella* there are only a few alar cells present, and this helps differentiate this genus from some other pendulous Hypnales mosses where the alar cells are more strongly differentiated.

RIGHT | *Orthostichella versicolor*, with its long pendant branches dangling from a tree branch in the Costa Rican rainforest.

DISTRIBUTION
Found in the tropical regions of America and Africa

ETYMOLOGY
Greek *orthos* = "straight" + *stichos* = "row," in reference to the arrangement of branch leaves

NUMBER OF SPECIES
9 accepted species

APPEARANCE
Primary stems creeping. Secondary stems erect or pendant, leaves spiraled, ovate, concave, costa mostly absent, leaf cells smooth and elongate, alar cells quadrate. Dioicous. Seta short, sometimes curved. Capsules ovoid to cylindric, with double peristome. Calyptra hood-shaped

HABITAT
Epiphytic on trunks or branches of trees

GLOSSARY

Acrocarpous: Mosses that produce **sporophytes** at the tips of the main stems.

Acuminate: Tapering gradually to a narrow point with slightly concave margins.

Alar cells: Specialized cells at the basal corners of a moss leaf that are distinct from the other leaf cells.

Antheridium (plural **antheridia**): The male reproductive organ of a bryophyte.

Apical: Relating to a structure's apex or tip.

Archegoniophore: In some **thalloid** liverworts, a tall stalk that bears the female reproductive organs.

Archegonium (plural **archegonia**): The female reproductive organ of a bryophyte.

Arthrodontous: Peristome teeth that are formed from parts of dead cells.

Awn: A bristle extending from the tip of a moss leaf formed from the leaf **costa** and protruding beyond the leaf.

Axil: The angle between a stem and a leaf or branch growing out of that stem.

Calcicolous: A plant that grows on a lime-rich **substratum**.

Calyptra (plural **calyptrae**): A small cap or hood formed from **gametophyte** tissue that covers the tip of the moss **sporophyte** as it develops.

Carpocephalum (plural **carpocephala**): In some **thalloid** liverworts, the disc-like receptacle that bears the female reproductive organs.

Chlorocyst: Green cells containing chlorophyll that occur among **hyaline** cells (**leucocysts**) in the mosses *Sphagnum* and Leucobryaceae.

Chlorophyllose: Having chlorophyll and being therefore green.

Chloroplast: An organelle within the plant cell that contains chlorophyll and within which photosynthesis takes place.

Cilia (adjective **ciliate**): Fine hair-like structures fringing a margin or surface.

Costa (plural **costae**): The thickened midrib or nerve of a moss leaf.

Cyanobacterium (plural **cyanobacteria**): Microorganisms closely related to bacteria that are capable of photosynthesis.

Decurrent: Extending below the point of origin on a structure.

Dehisce: The splitting or rupturing open of capsules to release spores.

Dioicous: With male and female reproductive organs produced on separate plants.

Diploid: When an organism contains two complete sets of chromosomes in the nuclei of its cells.

Distichous: Arranged in two rows.

Elaters: Sterile, spiraled cells interspersed among the spores of the liverwort capsule with a function to help disperse spores with their hydroscopic movement.

Endemic: The restriction of a specific organism to a particular region.

Endophyte (adjective **endophytic**): A fungus or other organism that lives within a plant for at least part of its life cycle.

Endosporic germination: When cell division within a spore begins before the spore wall ruptures.

Endostome: The inner ring of moss peristome teeth.

Endosymbiont: An organism that lives within the body of another organism for at least part of its life cycle.

Epiphragm: A circular membrane attached to the ends of the peristome teeth in the Polytrichales mosses.

Epiphyll (adjective **epiphyllous**): A plant that grows on the living leaves of another plant.

Epiphyte (adjective **epiphytic**): A plant growing on another plant.

Excurrent: Referring to the **costa** of a leaf when it extends beyond the leaf apex.

Exostome: The outer ring of moss peristome teeth.

Fibrils: Spiraled thickenings on the **hyaline** cells (**leucocysts**) in species of *Sphagnum*.

Flagelliform: A whip-like structure that is long, slender, and flexible.

Gamete: A **haploid** sex cell, either an egg or a sperm.

Gametophore: The leafy stem of a **gametophyte**.

Gametophyte: The **haploid**, multicellular, **gamete**-producing generation that in bryophytes is the dominant phase of the life cycle.

Gemma (plural **gemmae**): A vegetative **propagule**, usually formed of only a few cells and sometimes developing from specialized structures on the plant.

Haploid: When an organism contains only one complete set of chromosomes in the nuclei of its cells.

Hyaline: Colorless, lacking chlorophyll or other pigments.

Hydroids: Specialized narrow, empty cells that can conduct water and are found in some mosses such as the Polytrichales.

Hydroscopic: Readily absorbing moisture from the air.

Incubous: A descriptor for how leaves are attached to the stem when the upper edge of the leaf overlaps the margin of the leaf higher up on the stem.

Involucre: A tube of thallus tissue that protects the developing **archegonia** of some hornworts and thalloid liverworts.

Lamella (plural **lamellae**): A wall-like ribbon of leaf tissue running along the center of a leaf in some of the Polytrichales mosses.

Lamina (plural **laminae**): The flat blade of a moss leaf, not including the **costa** if present.

Leptoids: Specialized cells that can conduct the products of photosynthesis around the plant and are found in some mosses, most notably the Polytrichales.

Leucocysts: Large, empty, **hyaline** cells in the leaves of *Sphagnum* and some Leucobryaceae mosses.

Lignify: To convert to woody tissue by the deposition of **lignin** in the cell walls.

Lignin: A complex organic polymer that can provide structural support in plant tissues.

Lobule: The smaller lobe of a bilobed liverwort leaf, sometimes forming a sac-like structure.

Meiosis: A type of cell division where the DNA content of the resulting cells is halved.

Monoicous: With male and female reproductive organs produced on the same plant.

Monophyletic group: A single common ancestor and all of its descendants.

MYA: Abbreviation for millions of years ago.

Mycorrhizal fungi: Fungi that have a **symbiotic** association with a plant's root system.

Nematodontous: Peristome teeth that are formed from whole dead cells.

Ocellus (plural **ocelli**): A specialized cell in the leaves of some leafy liverworts that has a very large oil body and lacks chloroplasts.

Oil body: A membrane-bound organelle containing **terpenoids**, found in the cells of liverworts.

Operculum: A lid covering the mouth of the capsule that falls away when the spores are mature.

Papilla (plural **papillae**): A small, solid protuberance on a cell surface; can be variously rounded, branched, or C-shaped.

Papillose: Cell walls roughened by **papillae**.

Perianth: In leafy liverworts, a tube-like structure formed of leaves fused together that protects the developing **sporophyte**.

Perichaetium (plural **perichaetia**): The female sex organs plus a cluster of modified leaves surrounding them.

Perigynium (plural **perigynia**): In some thalloid liverworts, a swollen fleshy sleeve derived from the thallus tissue that surrounds the developing **sporophyte**.

Peristome: A single or double ring of teeth inside the mouth of a moss capsule.

Pinnate: A form of branching where branches are arranged either side of a main stem, like a feather.

Pleurocarpous: Mosses that produce sporophytes laterally along the branches rather than from the tips of the main stems.

Plicate: Pleated or furrowed.

Propagule: A structure, such as a **gemma** or **tuber**, that reproduces a plant vegetatively.

Protonema: The first growth stage of a bryophyte following germination of a spore, usually filamentous.

Pseudoelaters: In some hornworts, clusters of cells within the capsule that aid spore dispersal, developmentally distinct from **elaters** found in liverworts.

Pseudopodium: A **seta**-like structure found in the moss genera *Andreaea* and *Sphagnum* that is derived from the **gametophore** tissue.

Pyrenoid: Structures in the chloroplast of hornworts and some algal groups where starch is synthesized.

Rhizoid (plural **rhizoids**): Filaments arising from the stem or thallus of a bryophyte that have a role in anchoring the plant.

Seta (plural **setae**): The stalk of a moss or liverwort **sporophyte**.

Sinuose: Having a wavy margin.

Sporophyte: The **diploid**, spore-producing generation of the bryophyte life cycle.

Stereids or **stereid cells**: Thick -walled cells that support the leaves and stems of some mosses.

Substratum (plural **substrata**): Whatever an organism is growing on.

Succubous: A descriptor for how leaves are attached to the stem when the upper edge of the leaf is covered by the margin of the leaf higher up on the stem.

Symbiotic: A mutually beneficial association between two or more species where at least one of these species lives in or on another.

Teniola: The intramarginal border of linear, hyaline cells that characterize the leaves of *Calymperes* mosses.

Terpenoids: A group of organic chemical compounds.

Thalloid (or **thallose**): A flat plate or sheet of plant tissue.

Tuber: A non-photosynthetic **propagule** usually formed underground arising from **rhizoids**.

Vitta (plural **vittae**): A distinct strip in the leaves of some liverworts formed by enlarged cells running along the middle of a leaf or lobe.

FURTHER READING

BOOKS

Goffinet, B., & A. J. Shaw. 2009
Bryophyte Biology (2nd edn). Cambridge
University Press.

Vanderpoorten, A., & B. Goffinet. 2009
Introduction to Bryophytes. Cambridge University Press.

Bell, N. 2023
The Hidden World of Mosses. Royal Botanic Garden
Edinburgh.

Glime, J. 2021
Bryophyte Ecology (e-book: https://digitalcommons.mtu.
edu/oabooks/4).

FIELD GUIDES

There are a wealth of excellent bryophyte field guides
and floras available that tend to focus on specific regions
of the world. Two good examples are listed here:

Pope, R. H. 2016
*Mosses, Liverworts, and Hornworts:
A Field Guide to Common Bryophytes of the Northeast*.
Cornell University Press.

Atherton, I., S. Bosanquet & M. Lawley. 2010
*Mosses and Liverworts of Britain and Ireland:
A Field Guide*. British Bryological Society.

ONLINE IMAGE RESOURCES

Michael Lüth's website with habitat photographs
and photomicrographs of European mosses:
www.milueth.de/Moose/index.htm

Bryophyte images by Des Callaghan:
https://bryophytes.myportfolio.com/

CLASSIFICATION SOURCES

Brinda, J. C. & J. J. Atwood. October 23, 2024
The Bryophyte Nomenclator *www.bryonames.org*

Söderström, L. *et al.* 2016
"World checklist of hornworts and liverworts,"
PhytoKeys 59, 1–821

Goffinet B. & W. R. Buck.
Classification of the Bryophyta
*http://bryology.uconn.edu/
classification*

Bechteler, J. *et al.* 2023
"Comprehensive phylogenomic time tree
of bryophytes reveals deep relationships
and uncovers gene incongruences in the
last 500 million years of diversification,"
American Journal of Botany 110/11, 1–20

BRYOLOGICAL SOCIETIES

A number of bryological societies exist around the
world, some examples are listed below:

The International Association of Bryologists (IAB)
https://bryology.org/

The American Bryological and Lichenological Society
(ABLS) *https://www.abls.org*

The British Bryological Society (BBS)
www.britishbryologicalsociety.org.uk/

The Bryological–Lichenological Working
Group for Central Europe (BLAM)
https://blam-bl.de/

INDEX

Page numbers in *italics* refer to illustration captions.

PICTURE CREDITS

The publisher would like to thank the following for permission to reproduce copyright material: T = Top; B = Bottom; L = Left; R = Right; C = Center

All cover images by **Des Callaghan**, apart from:
Štěpán Koval: front cover (bottom, centre)
Shutterstock/Mykola Borduzhak: back cover

Alamy/Biosphoto: 228; Blickwinkel: 126; Sonja Blom: 27; Lee Rentz: 186; Studio75: 73T; Steve Taylor ARPS: 97T; David Whitaker: 109BL. © **Paul Bell-Butler**: 54. © **Lisa Bennett**: 65. © **Matt Berger**: 162, 163T. © **Nicola van Berkel**: 221L. **Andi Cairns**: 117T, 117B. **Des Callaghan**: 6, 7, 25L, 44, 49B, 51, 53, 63B, 67B, 69, 72, 74, 75T, 77, 79T, 80, 81, 83T, 84L, 86R, 87T, 87B, 89TL, 91T, 95B, 99BL, 100, 101T, 102, 103, 107BL, 109BR, 111T, 111B, 112, 117T, 118L, 118R, 119TL, 125T, 129, 132, 133B, 134, 137T, 138, 142, 153T, 154, 155T, 156, 157T, 160, 165T, 166, 167B, 169, 171, 173, 181T, 185T, 190, 193T, 199T, 199B, 202, 203, 210, 214, 215B, 219B, 225T, 227B, 229TL, 230. **Images by J G Duckett**: 64, 65T. **David Eickhoff**: 108. **Courtesy of Kathrin Feldberg**, University of Göttingen, Germany: 11T, 11BR. **Raita Futo**: 104. **Georgiasteel**: 212. **Getty Images**/B.S.P.I: 31; Little Dinosaur: 206. © **Stefan Gey**: 66. © **Nick Goldwater**: 213T. **Claire Halpin**: 25R, 40, 78, 85T, 89B, 91B, 92, 95T, 105T, 105B, 107BR, 109T, 127T, 139B, 147T, 148, 149T, 149B, 158, 170B, 174, 185B, 187T, 189L, 189R, 200, 201TL, 201BL, 208, 209T, 220, 221R, 223B. **Charles Hipkin**: 216, 217T. **Courtesy of Michael Ignatov**: 122, 123T, 123B, 150, 151T, 151B. © **Hugo Innes**: 75B, 163B. **L Jensen**, University of Auckland, N.Z.: 60, 96, 213B. © **Nina Kerr**: 61, 135T, 135B. **John Keys**: 67T. **Štěpán Koval**: 39, 50, 59, 63T, 68, 83BL, 83BR, 84R, 88, 106, 121B, 130L, 130R, 131, 133T, 136, 137B, 141T, 141B, 143T, 146, 147B, 159, 161T, 168, 170T, 182, 192, 194, 195L, 197T, 197B, 204, 205, 218, 219T. **Peter de Lange**: 97B. **Xuedong Li**: 117B. Michael Lüth: 37, 49T, 76, 90, 93B, 98, 99T, 107T, 121T, 139T, 145TL, 157B, 164, 167T, 191L, 193B, 222, 223T, 227T. **Dr David Meagher, courtesy of Dr Janine Major**: 176. **Missouri Botanical Garden**: 177T, 177B. **National Archives and Records Administration**, Washington, DC: 33B. © **National Taiwan Museum**: 181B. **Nature Picture Library**/Stephen Dalton: 23; Guy Edwardes: 2, 28-29; Orsolya Haarberg: 8-9; Alex Hyde: 71, 73B; Chien Lee: 115; Chris Mattison: 70; Duncan Mcewan: 195R; Yva Momatiuk & John Eastcott: 36; David Noton: 26-27; Nick Upton: 22; Staffan Widstrand: 183. © **Nils Nelson**: 152. **Kjell Nilsen**: 52. **Felipe Osorio**, Universidad Austral de Chile, Facultad de Ciencias Forestales & Recursos Naturales. Valdivia, Chile: 165B. **Sharon Pilkington**: 184, 196, 217B. **Randal**: 116, 144, 153B. **Hermann Schachner**: 145TR. **Alexander R. Schmidt**, University of Göttingen, Germany © 2014 Elsevier Ltd. All rights reserved: 11BL. **Science Photo Library**/Eye of Science: 18; Power and Syred: 117TR; Madga Turzanska: 15. **Shutterstock**/F-Focus by Mati Kose: 35; Hank Asia: 207. © **Pablo Silva**: 215T. **SLU Swedish Species Information Centre**/Malin Birgersson: 187B; Tomas Hallingbäck: 5, 125B, 140, 143B, 155B, 161B, 172, 191R, 201R, 209B; Niklas Lönnell: 180; Christopher Reisborg: 21, 48, 58, 62, 82, 85B, 86L, 89TR, 94, 99BR, 101B, 119B, 120, 124, 127B, 188, 198, 226, 229TR, 229B. **Société québécoise de bryologie**/Jean Faubert: 145B. **Image by Lil Stevens**: 24. **John Tann**: 128. Photos by Juan Carlos Villarreal A.: 46, 47T, 47B. **Joanna Wilbraham**: 9, 42L, 42R, 55, 79B, 93T, 110, 175, 224, 225B. © **Dan Wrench**: 211. **Scott Zona**: 178, 179. © **Zookanthos**: 231.

All reasonable efforts have been made to trace copyright holders and to obtain their permission for the use of copyright material. The publisher apologises for any errors or omissions and will gratefully incorporate any corrections in future reprints if notified.

ACKNOWLEDGMENTS

Research for this book relied upon the vast body of literature produced by bryologists around the world who are devoted to the study of these tiny, fascinating plants. I am extremely thankful to all the photographers who have contributed images. The standard of bryophyte photography has hit dazzling levels of excellence in recent years, allowing the intricate beauty of these plants to reach a much wider audience.

This book would not have been possible without the many hands that helped along the way. Jacqui Sayers first proposed the project with the vision that bryophytes could hold their own for 240 pages. Thanks go to Ruth Patrick for steering the journey; Katie Greenwood who sourced and organised the hundreds of images featured here; the team at Bright Press, Nick Pierce and Izzie Hewitt for editorial guidance; and Kevin Knight for fashioning the design.

I am grateful to the Natural History Museum, London, for supporting my involvement in this project and to my fantastic colleagues in the herbarium.

To Julie and Rowan, thanks for putting up with all the mossing.